# SOUND-FIELD FM AMPLIFICATION
## THEORY AND
## PRACTICAL APPLICATIONS

# SOUND-FIELD FM AMPLIFICATION

## THEORY AND
## PRACTICAL APPLICATIONS

**Carl C. Crandell,** Ph.D.
Department of Communication Processes and Disorders
University of Florida

**Joseph J. Smaldino,** Ph.D.
Department of Communicative Disorders
University of Northern Iowa

**Carol Flexer,** Ph.D.
School of Communicative Disorders
University of Akron

**SINGULAR PUBLISHING GROUP, INC.**
San Diego, California

| A SINGULAR AUDIOLOGY TEXT |
| Jeffrey L. Danhauer, Ph.D. |
| Audiology Editor |

**Singular Publishing Group, Inc.**
4284 41st Street
San Diego, California 92105-1197

©1995 by Singular Publishing Group, Inc.

Typeset in 10/12 Bookman by So Cal Graphics
Printed in the United States of America by McNaughton & Gunn

*All rights, including that of translation reserved. No part of this publication may be reproduced, stored in a retrieval system, or transmitted in any form or by any means, electronic, mechanical, recording, or otherwise, without the prior written permission of the publisher.*

**Library of Congress Cataloging-in-Publication Data**
Crandell, Carl C.
    Sound-field FM amplification: theory and practical applications/
  Carl C. Crandell, Joseph J. Smaldino, Carol Flexer
       p.   cm.
  Includes bibliographical references (p. ) and index.
  ISBN 1-56593-450-4
    1. School buildings—United States—Acoustics.  2. Auditory
perception—United States—Technological innovations.  3. Classroom
management—United States.  I. Smaldino, Joseph J.  II. Flexer,
Carol Ann.  III. Title.
LB3241.5.C73   1995
371.6'2—dc20
                                          94–45294
                                          CIP

# CONTENTS

# FOREWORD

My poor performance in several statistics courses was the closest I came to flunking out of my doctoral program. I just couldn't get it—not necessarily the material, but its classroom presentation. The professor would greet us when he entered the classroom, and that may have been the last time we saw his face for the entire period (a little overstated, but not much). He would face the blackboard, develop and write the various formulae on the board, explaining them to the blackboard as he went along. When he finished lecturing to the blackboard he would turn around and ask if there were any questions. Now I expected more difficulty than the other students because of my hearing problem, though I certainly didn't expect an almost complete absence of comprehensible speech. What I didn't realize at the time, but did later, was that I was not the only one in the class to have hearing problems, but these were experienced by normal-hearing students and not just from someone with a hearing loss. These students also felt that if they could have heard the professor better, they would have had an easier time learning the material. This was at a doctoral program at a prestigious university in a quiet classroom composed of intelligent and highly motivated young men and women—nobody with learning problems, attention deficit disorders, minimal hearing losses, or what have you. How much more this observation applies in other educational settings, and what can be done about it, is the subject of this book by Carl Crandell, Joseph Smaldino, and Carol Flexer.

Unlike a university classroom, noise is a ubiquitous presence in most other classrooms. The students are an unselected mix of special and regular students, manifesting the usual range of performance, intelligence, maturity, and neurological status. All of them, however, as Carol Flexer never seems to tire of telling us, have to *hear* in order to *learn*. And within broad, and not too well defined limits, the better they hear, the more they are going to learn. Some children, with completely intact central auditory nervous systems, have a relatively easy job of extracting speech from a background of noise; others require a more favorable speech-to-noise ratio to compensate for central or peripheral auditory difficulties. Hearing the teacher, however, is necessary for all of them.

It seems so obvious. Put a child in a position where he or she cannot hear the classroom lessons and the child will not learn them. Put such a child in a position where only some portion of the signal is clearly, and inconsistently, audible and the child's performance will be variable. Some children will do well; some may do nearly or just as well, but will have to expend a great deal more effort; some will get by using compensatory strategies and cues and a highly motivated approach to learning; some will find their attention wavering after a few hours and "tune-out" and others will fall by the wayside, being unable to integrate the acoustic and linguistic fragments they hear completely into a meaningful gestalt. In other words, **some** children can benefit **all** the time from a more audible speech signal from the teacher, but likely **all** the children can benefit **some** of the time.

As I said, it seems so obvious, yet just several hours before I started to write these words, I received a phone call which illustrates how difficult it is to communicate such a simple concept. The call was from the principal of a denominational Sunday School, the kind of place not generally reputed as a quiet and serene learning environment. She had a number of special needs children in her school and, as a responsible professional, she wanted to be sure that she was responding to their needs. None of the children wore hearing aids, and her questions to me related to my management suggestions for the children with learning disorders and attention deficits. I explained that this was not my area of expertise, but that I was sure that she could help these children, and many of the others as well, if she amplified the classroom. As I explained the potential advantages, I could sense her skepticism and disbelief. She was asking me to recommend learning strategies for special needs children, and I was countering with suggestions that an improved speech-to-noise ratio could help! This was an intelligent, committed, and sensitive professional who really wanted to respond to the individual needs of her children, but who could not grasp the simple relationship between hearing and learning.

This is precisely the gap this book is designed to fill. Finally, we have a single source which covers both the theoretical foundations for the use of sound-field amplification in the classroom, and the practical application of the concept. The results with children are reported as well as the enthusiastic endorsement by teachers. As a book, a published and public item, the concept can now take on academic respectability and credibility not possible with my simple assertions over the telephone. In that sense alone, this book is going to fill a real need. If I could have referred the principal to this book, my recommendations would have taken on increased legitimacy.

Sound-field FM amplification is not a panacea for all the educational problems occurring in the myriad of classrooms in our country, nor is it being presented as such. If expectations are too high, disappointment is sure to come. Children will still act like kids; some will pay less attention to the lessons than others; some are more able than others; and noise will still be an occasional interfering presence. The device is an *additional* tool for learning. It will make the teaching and learning tasks easier, but not effortless. I doubt if it will do away with homework! Still, the educational achievements and classroom behavior in our schools is not such that we can afford to ignore any device that can offer significant help in such an effortless manner. All the teacher has to do is what she or he has been doing (i.e., talk), and the amplification device does the rest.

As we have seen for many years, however, in respect to the individual use of FM systems (and personal hearing aids) by hearing-impaired children in schools, Murphy's law ("anything that can go wrong, will") is not going to be repealed as it relates to sound-field FM classroom amplification. Teachers have to be sufficiently convinced of the merits of this approach to deal with the occasional glitches and frustrations. And that's where this book comes in. Not only as motivational document, but armed with the information contained in the book, many problems can be avoided and others solved as they occur. It is, in brief, a source document, the first if its kind, for a proved, effective, and relatively inexpensive educational tool. Carl Crandell, Joseph Smaldino, and Carol Flexer were among the pioneers in using and recognizing the merits of this tool, and they are to be congratulated for presenting it in such a comprehensive fashion to the rest of us.

Mark Ross, Ph.D.
Professor Emeritus
University of Connecticut

# PREFACE

## PURPOSE OF THE BOOK

The purpose of this book is to provide a comprehensive and cohesive guide for the use of small, frequency modulated sound-field FM amplification systems. This book addresses both theoretical and practical issues with an emphasis on application of information to real world situations. Worksheets and checklists are included at the end of many chapters.

## AUDIENCE

This book is intended for the large number of individuals who might use and/or benefit from sound-field FM amplification systems, including audiologists, speech-language pathologists, school administrators, classroom teachers, special education teachers, teachers of children who are hearing impaired, parent-teacher associations, parents, civic groups, and equipment manufacturers.

## BOOK CONTENT

There are two general sections of this book: Theoretical foundations for the use of sound-field FM amplification, and practical applications of sound-field FM amplification. The reason for these divisions is that people who are trying to obtain sound-field FM systems for their schools first need to show that there is a need for these systems. There is a research basis for amplifying regular classrooms as well as special education classrooms. Next, interested persons will need to know how to select, install, use, and maintain this equipment. This book has been designed to answer all of these questions.

There seems to be a general misconception that the child, the classroom, and the teacher are "constants." That is, the belief is that these items stay about the same from day to day. In fact, the child, classroom, and teacher are all "variables" in the educational process. Young children's hearing, speech perceptual ability, attention, motivation, position in the classroom, and interest may vary day to day, or even during the same day. The classroom environment varies as noise levels inside the room, outside the building, and outside the classroom change. The teacher's energy, voice level, interest, motivation, preparation, and position in the classroom often vary throughout the day. The point is, when students, classrooms, and teachers are recognized as variables in the learning process and not as constants, we are compelled to understand and manage them. To let them vary randomly throughout the day is to undermine students' opportunities to learn in the classroom.

Accordingly, Chapter 1 reviews the rationale for using sound-field amplification systems in the classroom. Chapter 2 discusses the child/student as a variable in the learning process. Data will be presented that show that children cannot be viewed as small adults relative to speech perception. Chapter 3 elaborates on the classroom environment as a variable in the educational process. Chapter 4 presents an overview of the populations of children who have more difficulty learning in a classroom environment than do typical children.

The practical applications portion of the book begins with Chapter 5, and describes acoustic measurements of the classroom. Acoustic goals are set for the classroom and pre- and post-measurements are completed to determine the effectiveness of those acoustical treatments. Chapter 6 examines various procedures for modifying the classroom environment. Chapter 7 describes the process of identifying and managing teaching style. The teacher is a key player in the scenario of effectively utilizing sound-field equipment. Chapter 8 summarizes the available literature concerning sound-field amplification. The nuts and bolts of actually amplifying the classroom is presented in Chapter 9. An acoustic goal has been set. Now that the classroom has been made quieter and the teacher and teaching style have been identified and managed, how do we address the residual? How do we achieve the goal of improving the speech-to-noise (S/N) ratio by 10 dB throughout the learning area(s)?

Chapter 10 presents topics and strategies for inservicing the classroom teacher. The teacher is a key player. He or she must be comfortable with the equipment and know how to use it effectively. Chapter 11 discusses the critical issue of the development of listening skills. How can listening strategies enable teachers and pupils to make maximum use of sound-field technology? Chapter 12 presents many commonsense tips for marketing sound-field amplification systems. How can we obtain funding for units? How can we show educators that the equipment is worthwhile? Is sound-field amplification cost effective?

Sound-field technology offers exciting promise for educating children. Larry Bracken, a retired School Superintendent, said, "If kids hear better, they do better in school! I feel that the promotion and use of sound-field FM equipment is the single most innovative and important project for children with which I have been associated!"

# DEDICATION

To the students born and unborn who by oversight of neglect suffer the consequences of inappropriate classroom acoustics.

# CONTRIBUTORS

**Laurie A. Allen, M.A.,**
Educational Audiologist
Keystone Area Education Agency
Dubuque, Iowa

**Karen Anderson, Ed.S.**
Educational Audiologist
Rainier Audiology Consulting
   Services
Puyallup, Washington

**Patricia Blake-Rahter, M.S.**
Audiologist/Clinical Instructor
University of South Florida
Communication Sciences
Tampa, Florida

**Carl C. Crandell, Ph.D.**
Assistant Professor—Audiology
Department of Communication
   Processes & Disorders
University of Florida
Gainesville, Florida

**Carolyn Edwards, M.Cl.Sc., M.B.A.**
Director, Auditory Management
   Services
Toronto, Ontario
Canada

**Carol Flexer, Ph.D.**
Professor of Audiology
School of Communicative Disorders
University of Akron
Akron, Ohio

**Gail Gegg Rosenberg, M.S.**
Audiologist/Program Specialist
School Board of Sarasota County
Sarasota, Florida

**Mark Ross, Ph.D.**
Professor Emeritus
University of Connecticut
Storrs, Connecticut

**Joseph J. Smaldino, Ph.D.**
Professor of Audiology
Head, Department of
   Communicative Disorders
University of Northern Iowa
Cedar Falls, Iowa

# ACKNOWLEDGMENTS

**Carl Crandell** would like to sincerely thank the following individuals: (1) my dear wife, Lorraine, and beautiful daughter, Danielle, for putting up with me during the writing, editing, and proofing of this book. Thanks for the quiet times, guys; (2) Joe Smaldino and Carol Flexer for being greater friends and colleagues. It was an honor working with both of you on this "labor of love"; (3) all those at Singular (including Sadanand, Angie, Jeff, Marie, and Randy) for allowing this book to become a reality; (4) each of the authors who contributed chapters to this book. It was a real pleasure working with each of you; (5) Mark Ross, who wrote the greatest Foreword I've ever read; (6) Alyssa Needleman and Georgeanne Self for reading many earlier drafts of this book; and (7) last, but certainly not least, the good Lord for giving me this wonderful opportunity.

**Joseph Smaldino** would like to thank Frederick Berg, James Blair, Steve Viehweg, and all of the faculty at Utah State University whose foresight and skill in obtaining the Listening in the Classroom Project Grant, stimulated and brought together the primary authors; and the Smaldino family, especially Sharon for her support and critical thinking of the manuscript.

**Carol Flexer** would like to express gratitude and appreciation to Carl Crandell and Joe Smaldino. Their knowledge, dedication, and hard work have made this book a reality! A special acknowledgment needs to be extended to Mark Ross. He has been and continues to be an inspiration to me. And, finally, I would like to thank my husband Pete, my children Heather, Hillari, and David, and my new little grandson Yehuda, for their consistent and patient support.

From Carol and Joseph . . . Carl "you did good."

PART

A

# A THEORETICAL FOUNDATION FOR THE USE OF SOUND-FIELD FM AMPLIFICATION

# CHAPTER 1

# RATIONALE FOR THE USE OF SOUND-FIELD FM AMPLIFICATION SYSTEMS IN CLASSROOMS

*Carol Flexer*

The room has linoleum floors, cinderblock walls, cement ceiling, bare windows, and chalk boards on the front and back walls. The windows are open for ventilation and a large fan is roaring in the back corner of the room. There is a busy street adjacent to the building and a lawn is being mowed. The door to the hallway is open and sounds from other rooms, passing feet, and the gym are pouring into the room. Thirty-one restless six-year-old children are sitting at desks. Some of the children have fluid in their middle ears, some do not speak English as their first language, some have weak attending skills, some have language and/or articulation problems, and some are just typical children. The single adult in the room is fatigued from trying to maintain order and attention. This is a typical classroom.

Adults would not tolerate the above acoustic environment for very long. Yet we place millions of children in this situation for at least six hours a day, year in and year out. Then we are surprised when they do not learn at the pace or to the degree that is expected. Mainstream classrooms are auditory-verbal environments; listening is the primary modality for learning. Instruction is presented through the speech of the teacher with the underlying assumption that pupils can hear clearly and attend to spoken communication. To the extent that students cannot consistently and clearly hear the teacher, the entire premise of the educational system is undermined. Ironically, the acoustic environment is rarely considered when designing a school.

Elliott, Hammer, and Scholl (1989) tested children with normal hearing to determine their ability to perform fine-grained auditory discrimination tasks such as hearing "pa" and "ba" as different syllables. The authors classified 80 percent of the primary-level children in their study either as progressing normally or as having language-learning difficulties. Findings such as these suggest that auditory discrimination is associated with the development of basic academic competencies that are essential for success in school. The young child who cannot detect and recognize phonemic distinctions is at extreme risk for academic failure. Therefore, a logical first step in facilitating the educational process is managing and enhancing the acoustic learning environment.

The ability to discriminate individual phonemes, to perceive word-sound differences, is defined as *intelligibility*. This is distinguished from *audibility*, which is simply the ability to detect the presence or absence of speech. If, because of poor (or even typical) classroom acoustics, weak attending skills, language or articulation problems, ear infections, and so on, a child cannot discriminate the word *fix* from *six*, for example, he or she will not learn appropriate semantic distinctions unless deliberate intervention occurs.

This book is about understanding the critical nature of classroom listening/learning, and about managing and enhancing that environment. Specifically, this book is about a technology that has been designed to provide accessibility to a teacher's spoken instruction: sound-field FM amplification systems.

## DESCRIPTION AND PURPOSE OF SOUND-FIELD FM SYSTEMS

Frequency-modulated sound-field FM amplification systems are similar to small, high fidelity, wireless, public address systems that are self-contained in a classroom. Specifically, the teacher wears a small, unobtrusive wireless microphone; thus, teacher mobility is not restricted. His or her speech is frequency modulated onto a carrier wave that is sent from the transmitter to the receiver where it is demodulated and delivered to the students through one to five wall-or-ceiling-mounted loudspeakers. (See Figures 1–1a, b, c, d, and e for pictures of several sound-field systems.)

The purpose of the equipment is to amplify the teacher's voice throughout the classroom, thereby providing a clear and consistent signal to all pupils in the room no matter where they or the teacher are located. The positioning of the remote microphone close to the mouth of the teacher or other desired sound source creates a favorable speech-to-noise ratio (S/N ratio). The more favorable the S/N ratio, the clearer and more intelligible the speech signal that is received by the pupils in the classroom. Therefore, sound-field FM amplification can, if properly adjusted, counteract weak teacher voice levels and background noise by increasing the overall speech level, substantially improving the S/N ratio, and

1–1a

**Figure 1–1.** A sound-field amplification system is comprised of a wireless teacher-worn microphone/transmitter, an amplifier, and one to four loudspeakers positioned around the room or in the ceiling; five different sound-field units are shown in 1a to 1e. (1–1a is Courtesy of Lifeline), (1–1b is Courtesy of Audio Enhancement), (1–1c is Courtesy of Audio Enhancement, (1–1d is Courtesy of Anchor Audio), (1–1e is Courtesy of Telex) *(continued).*

**1–1b**

**1–1c**

**Figure 1–1** *(continued)*

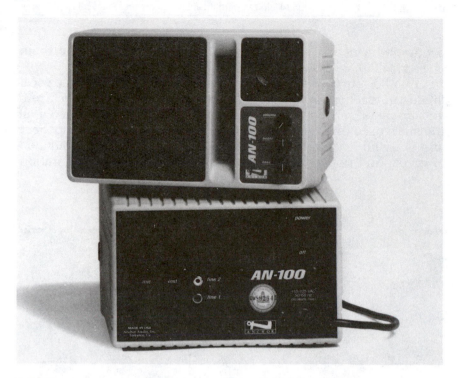

1–1d

1–1e

producing a nearly uniform speech level throughout the room (Flexer, Millin, & Brown, 1990). (See Figures 1–2a and b.)

Another purpose of sound-field amplification is to reduce vocal fatigue in teachers. A number of professionals, including lecturers and teachers, rely on voice as a primary tool for interaction. Of all the skills used, the voice ranks as the primary speaking component and is a vital teaching aid. The teacher needs to use a combination of verbal and nonverbal skills in order to maximize presentation

**1–2a**

**Figure 1–2.** The logic of sound-field amplification is that it can provide a favorable and consistent speech-to-noise ratio for all pupils in the classroom no matter where they or the teacher are located (1–2a is Courtesy of Phonic Ear), (1–2b is Courtesy of Custom Audio Design).

1–2b

effectiveness. It is essential that a presentation be of adequate loudness. With an inaudible message, the learning process is in jeopardy. Even a mild voice disorder can result in reduced job effectiveness for those persons in occupations that place a high demand on vocalization. Vocal attrition has been reported as one of the most frequent problems that teachers experience. With vocal rest, symptoms can ameliorate and the voice may perceptually return to normal. However, most professionals cannot afford lengthy periods of time away from their job commitment. Therefore, using an alternative mode of communication, such as sound-field amplification, could assist in augmenting voice projection.

## COMPUTER ANALOGY

One way to understand the effect of sound-field FM amplification better is to use a computer analogy (Flexer, 1994). A student must have information/data in order to learn. The primary avenue for entering information into a student's brain in a typical classroom is through his or her ears via hearing. If data are entered incompletely, inaccurately, or inconsistently (analogous to using a malfunctioning computer keyboard or to having fingers on the wrong keys of a computer keyboard) the pupil will have faulty data to process. How can a student be expected to learn new concepts, vocabulary, and thinking skills when the information that reaches the brain is deficient? (See Figure 1–3.) Is it the student's fault that the entered data are deficient? Is the computer program in error if the keyboard entry is inaccurate?

Sound-field FM amplification provides a reliable keyboard for more consistent, complete, and controlled data entry. (See Figure 1–4.) Logic would suggest that if students received clearer and more intact instructional signals, they would have better opportunities to learn in the classroom.

## THE INVISIBLE ACOUSTIC FILTER
## EFFECT OF HEARING PROBLEMS

What typically happens if a child does not hear well? Asked another way, how can an infant's or child's development be compromised by unidentified and unmanaged hearing problems of any type and degree?

The "power" of hearing tends to be underestimated because hearing loss is invisible. The negative effects of hearing problems typically are associated with problems other than hearing impairment, because the effects of hearing loss are ambiguous and difficult to conceptualize (Davis, 1990; Ross, 1991). For example, when a young student responds inappropriately to instructions, cannot stay on topic, appears to daydream or to be distracted, or has poor language or reading skills, the cause of the child's behaviors may be attributed to slow learning or to noncompliance rather than to hearing problems.

The ambiguity of hearing problems is magnified by the tendency to categorize hearing loss into the dichotomous classifications of normally hearing or deaf (Ross & Calvert, 1984). Children with minimal, mild, or moderate hearing impairments obviously are not deaf, so their hearing problems are often erroneously thought to present only a minor barrier to classroom performance

**Figure 1–3.** Experiencing hearing problems in the classroom is like having a malfunctioning computer keyboard that enters incomplete, faulty, and inconsistent data to the brain. (Illustration by Josh Klynn: Reprinted with Permission from Thieme Medical Publisher, Inc., Flexer, 1995.)

**Figure 1–4.** A sound-field amplification system can facilitate data entry by providing a more accessible keyboard. (Illustration by Josh Klynn: Reprinted with Permission from Thieme Medical Publisher, Inc., Flexer, 1995.)

(Bess, 1985; Davis, 1990). In reality, there is no such thing as an "insignificant" hearing problem in children. Any hearing problem can obstruct a child's reception of spoken communication.

Davis (1990) and Ling (1986) describe hearing impairment as an invisible acoustic filter. The negative effects of the hearing problems are observed but the causal filter is hidden. The primary negative effect of the invisible acoustic filter of hearing impairment is its detrimental impact on spoken language development. If an infant or young child cannot hear, then he or she cannot learn to speak unless deliberate intervention occurs (Ling, 1989). If a young child does not hear clearly and consistently, spoken language skills will not be clear either unless thoughtful intervention occurs.

The secondary negative effect of the invisible acoustic filter of hearing problems is its interference with the higher-level linguistic skills of reading and writing. If a child does not hear clearly enough to identify word-sound distinctions, then clear spoken language will not develop. If spoken language skills are deficient, then reading skills likely will be deficient also because reading is built upon speaking (Simon, 1985). To continue with the progressive nature of the negative effects of the acoustic filter, if a child has poor reading skills, then academic options will be limited (Wallach & Butler, 1984). The cause of this entire unfortunate, yet largely preventable chain of events, is the ambiguous, invisible, underestimated and often unidentified acoustic filter effect of hearing problems. Until the first order event of hearing problems is recognized, identified, and managed, intervention at the secondary levels of spoken language, reading, and academics will likely be ineffective.

School personnel may misinterpret the cause of a student's behaviors that result from the acoustic filter effect of hearing problems. For example, a teacher recently said, "We all know that Johnny has chronic ear infections and hearing loss, but his problem is not hearing loss; his problem is that he does not pay attention and he has reading difficulties." Johnny's problem **is** his hearing loss; his hearing loss is negatively effecting his attention, language, and reading skills. Until Johnny's hearing is enhanced through technology to allow him to hear clearly and consistently and subsequently to learn word/sound distinctions, he may have limited success improving reading skills. **Hearing cannot be bypassed in the chain of intervention. Hearing is a first-order event for classroom learning.**

The relationship between hearing loss and academic failure is well documented (Berg, 1986, Bess, 1985; Brackett & Maxon, 1986; Davis, 1990; Ross & Giolas, 1978). Even a minimal hearing problem or a unilateral hearing loss can cause significant academic difficulty.

To return to the computer keyboard analogy, providing the most reliable and consistent keyboard is the necessary first step for the development of spoken language, reading, and academic competencies. The longer a child receives faulty data entry, the more deficient the data base. Faulty data entry can be caused by hearing problems, attention difficulties, language problems, and other factors.

What happens to all of the previously entered inaccurate and deficient information once a proficient keyboard is provided? Does all of the inaccurate information magically convert to complete data? Or does the missed and inaccurate information have to be reentered into the system? Remember, hearing is the crucial and necessary first step. Once hearing has been accessed, the student has an opportunity to learn language as the basis for acquiring knowledge of the world. All levels of the acoustic filter need to be understood to be accessed. The longer a child has inaccurate data entry, the more destructive will be the acoustic filter effect on the child's overall life development. Conversely, the more clear and complete the entered information, the better the student's opportunity to acquire spoken language, learn reading skills, and progress in the development of academic competencies.

## FEDERAL LAWS THAT GOVERN "ACOUSTIC ACCESSIBILITY" IN CLASSROOMS

Three Federal Laws mandate services for children with hearing problems in schools: Education for All Handicapped Children Act of 1975 (Public Law 94-142); Education of the Handicapped Act Amendments of 1986 (Public Law 99-457); and The Rehabilitation Act of 1973 (specifically, Section 504). In 1990, Public Law 99-457 was amended by Congress and its name changed to the Individuals with Disabilities Education Act, commonly referred to as IDEA; Education of the Handicapped Act Amendments of 1990.

IDEA provides educational accommodations for qualifying students by developing an IEP (Individualized Education Plan) and by using Special Education funds. Most students who will benefit from sound-field FM amplification systems have not performed poorly enough to qualify for these Special Education services. Therefore, Section 504 of the Rehabilitation Act of 1973 probably will be the most relevant legislation for them.

Section 504 protects children and adults with different kinds of mental and physical disabilities. The term disability is very broad and includes any person who (DuBow, Geer, & Strauss, 1992, p. 51):

1. has a physical or mental impairment that substantially limits one or more major life activities,
2. has a record of such an impairment, or
3. is regarded as having such an impairment.

Major life activities include the ability to take care of oneself, hearing, walking, doing manual tasks, speaking, seeing, breathing, learning, and working.

The concept that will enable audiologists to recommend classroom amplification systems for children in regular classrooms, under Section 504, is *acoustic accessibility*. We can advocate, proactively, that a child's hearing problem, language problem, or attention problem interferes with his or her opportunity to have access to spoken instruction. Therefore, that child is being denied an appropriate education. By performing speech-in-noise testing in a sound room, an audiologist can provide evidence that a child cannot hear clearly in typical classroom environments (Flexer, 1995). The sooner a child has access to a reliable and consistent keyboard, the better opportunity that child will have to receive an appropriate education. **It is never too early to identify and manage hearing problems.**

## CHECKLIST OF IMPORTANT POINTS ABOUT SOUND-FIELD AMPLIFICATION

Following is a checklist of important points that were discussed in this chapter.

- Hearing and listening form the invisible cornerstones of the educational system.
- Auditory discrimination is associated with the development of basic academic competencies that are essential for success in school.
- Hearing problems, whether minimal, fluctuating, or unilateral, if unmanaged can have a negative impact on the development

of spoken language communication, reading, writing, and academic competencies.

• Spoken instruction might be audible to someone with hearing or attention problems, but not necessarily intelligible enough to hear one word as distinct from another.

• Hearing is a primary way that information is entered into a student's brain in a classroom; data input precedes data processing.

• Sound-field FM technology offers a way to amplify classrooms through the use of one to five wall or ceiling-mounted loudspeakers; the teacher wears a wireless microphone transmitter.

• Sound-field technology allows all children in the classroom to benefit from an improved and consistent signal-to-noise ratio no matter where they or the teacher are positioned.

• Section 504 of the Rehabilitation Act of 1973 is the federal law that can assist audiologists in recommending sound-field FM amplification for children who do not have "acoustic accessibility" to typical classroom instruction.

# REFERENCES

Berg, F. S. (1986). Characteristics of the target population. In F. S. Berg, J. C. Blair, J. H. Viehweg, & A. Wilson-Vlotman (Eds.), *Educational audiology for the hard of hearing child* (pp. 1–24). New York: Grune & Stratton.

Bess, F. H. (1985). The minimally hearing-impaired child. *Ear and Hearing, 6*, 43–47.

Brackett, D., & Maxon, A. B. (1986). Device alternatives for the mainstreamed hearing-impaired child. *Language, Speech, and Hearing Services in Schools, 17*, 115–125.

Davis, J. (Ed.). (1990). *Our forgotten children: Hard-of-hearing pupils in the schools*. Bethesda, MD: Self Help for Hard of Hearing People.

DuBow, S., Geer, S., & Strauss, K. P. (1992). *Legal rights: The guide for deaf and hard of hearing people* (4th ed.). Washington, DC: Gallaudet University Press.

Education of Handicapped Children, P.L. 94-142 Regulations. (1977, August 23). *Federal Register, 42*(163), 42474–42518.

Education of the Handicapped Act Amendments of 1986, P.L. 99-457. (1986, October 8). *United States Statutes at Large, 100*, 1145–1177.

Education of the Handicapped Act Amendments of 1990, P.L. 101–476. (1990, October 30). *United States Statutes at Large, 104*, 1103–1151.

Elliott, L. L., Hammer, M. A., & Scholl, M. E. (1989). Fine grained auditory discrimination in normal children and children with language-learning problems. *Journal of Speech and Hearing Research, 32*, 112–119.

Flexer, C. (1994). *Facilitating hearing and listening in young children*. San Diego: Singular Publishing Group, Inc.

Flexer, C. (1995). Classroom amplification systems. In R. Roeser, and M. Downs, (Eds.). *Auditory disorders in school children* (3rd ed., pp. 235–260). New York: Thieme Medical Publishers, Inc.

Flexer, C., Millin, J. P., & Brown, L. (1990). Children with developmental disabilities: The effect of soundfield amplification on word identification. *Language, Speech and Hearing Services in Schools, 21*, 177–182.

Ling, D. (1986). On auditory learning. *Newsounds, 11*, 1.

Ling, D. (1989). *Foundations of spoken language for hearing impaired children*. Washington DC: The Alexander Graham Bell Association for the Deaf.

Rehabilitation Act of 1973, P.L. 93-112. (1973, September 26). *United States Statutes at Large, 87*, 355–394.

Ross, M. (1991). A future challenge: Educating the educators and public about hearing loss. In C. Flexer (Ed.). Current issues in the educational management of children with hearing loss. *Seminars in Hearing, 12*, 402–413.

Ross M., & Calvert, D. (1984). Semantics of deafness revisited: Total communication and the use and misuse of residual hearing. *Audiology, 9*, 127–145.

Ross, M., & Giolas, T. G. (Eds.). (1978). Auditory management of hearing-impaired children. Baltimore: University Park Press.

Simon, C. S. (1985). *Communication skills and classroom success*. San Diego, CA: College-Hill Press.

Wallach, G. P., & Butler, K. G. (Eds.). (1984). *Language learning disabilities in school age children*. Baltimore, MD: Williams & Wilkins.

# CHAPTER
# 2

# SPEECH-PERCEPTION PROCESSES IN CHILDREN

*Joseph Smaldino*

**M**uch of this book is devoted to a discussion of acoustic variables, such as signal-to-noise ratio and reverberation in the classroom. Indeed, the impact of these acoustic factors on a student's ability to perceive spoken language is our main thesis. In order to understand the impact of acoustic variables and the basis for acoustic management and habilitative intervention, a simple model of the speech perceptual process may be helpful. Using the model, this chapter will examine how the speech perceptual competency that a student brings into the classroom is an important variable, that determines how well a student can use acoustic speech information and attain teacher understanding.

**The primary goal of the classroom educational process is to share experiences, exchange ideas, and transmit knowledge**. This is accomplished by the students not only because they are able to receive specific acoustic information, but because they are able to relate these often ambiguous cues within the context of a language structure and use circumstantial cues (Denes & Pinson, 1993).

## SPEECH PERCEPTION MODEL

A simplified model of speech perception is shown in Figure 2–1. The central thesis of this model is that incoming and stored knowl-

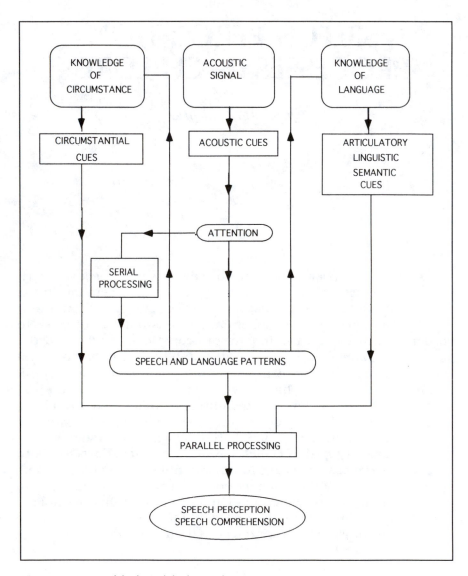

**Figure 2–1.** Simplified model of speech perception.

edge interact to result in accurate speech perception. Distortions of the incoming information and/or incomplete stored knowledge can impair a student's ability to perceive the speech information so critical to success in school. Let us examine how incoming and stored information interact differently to produce accurate and faulty speech perception.

In order for accurate speech perception to occur, stored knowledge bases must exist and be intact. The knowledge bases are developmental and become more complete and refined with experience. Normal principles of language acquisition operate to guide the induction of an acoustic/linguistic language system (McCormick & Schiefelbusch, 1984; Sanders, 1993). From Figure 2–1, an acoustic signal, containing acoustic cues critical to speech perception, attracts the attention of the developing language processing system. In the beginning of the language learning process much of this acoustic information is novel and requires close scrutiny. The signal is analytically processed in a mode called *serial* processing. It is during serial processing that speech and language acoustic patterns are analyzed, structured, and stored. An example of serial processing is when one first learns a foreign language. Each word must be broken down into individual sounds, rules for linking words must be studied, and vocabulary learned. Processing is slow and native speakers seem to be speaking at breathtakingly fast rates. Ability to understand what is being said is greatly influenced by distortions of the incoming signal, background noise, extent of visual cues, and speaker rate. Eventually, the acoustic patterns are linked to articulatory movements and linguistic and semantic rules; in addition, circumstantial cues such as identity of the speaker and visual information, are also tied to the patterns. All of these rules and cues work together to duplicate and constrain the possible interpretation of the incoming signal. This system is so efficient that, once formed, another more efficient form of attentional processing can be used. With so much duplication and constraint on the incoming signal, the signal no longer needs to be dissected in order to be understood; it can now be synthetically processed in larger chunks in a less cognitively demanding mode called *parallel* processing. An example of parallel processing is after a listener becomes familiar with a foreign language, accurate perception occurs rapidly and effortlessly. Native speakers no longer are speed talkers and one can keep up with the message they are transmitting. Distortion of the speech or background noise do not affect understanding as drastically. Normal speech perception can be thought of as such a process. It should be noted that while the incoming acoustic signal is an important variable, interaction with knowledge about the language and circumstantial cues are also important variables.

Inaccurate speech perception can occur because of a break-down in any of the aforementioned variables. Let us take each variable and describe possible effects from the model in Figure 2–1. The acoustic speech signal contains cues to accurate speech perception. If there is a hearing loss, even a slight hearing loss, some of these cues will be attenuated, filtered, or distorted. Poor classroom acoustics can have the same effect on the incoming signal and have similar effects on the speech processing model. Let us assume that an incoming signal is distorted in frequency, intensity, and time domains. The novel signal will be attended to by the student and serial processing will occur. As part of this processing mode, linkages to the other knowledge bases will be made. In a very young child, these distorted signals will form the knowledge bases, and they will be as a consequence incomplete. There is no need to dwell upon the well known faulty language systems which develop in most severely hearing-impaired children without intervention (Boothroyd; 1984; Sanders, 1993). It is logical, however, that subtle language linkage abnormalities will occur even if the distortion to the incoming signal is slight. In older children, who have normally established language linkages, distortion can make an incoming signal novel, that is, different from expectation based on already established language linkages. In either case, the child may be forced into a serial processing mode, an analytical mode, in an attempt to reconcile the discrepancies between the incoming signal and the language knowledge base. As noted before, the serial processing mode is slower and requires a great deal of cognitive energy to attend to and try to establish linkages to the knowledge bases. Often these children cannot keep up with the speed of the incoming signal and need more time to reconcile the incoming signal to what they know. Remember the beginning foreign language learning example. Becoming lost in the incoming information flow, the child may only receive fragments of the signal, or lose attentiveness and process unimportant parts of the signal. The parallel processing mode is never fully engaged. Further degradation of the incoming signal due to noise or other distortion makes the situation worse because the child has little reliable redundant information to fall back on. Such is faulty speech perception.

So far, this chapter has demonstrated that the analysis, structure, and storing of the incoming signal necessary for speech perception is a multi-variable process. Not only is it dependent upon receipt of specific acoustic information, but also upon the ability to relate these often ambiguous cues within the context of a language structure and use of circumstantial cues. Breakdowns in the system produce faulty speech perception. Much of this book is devoted to breakdowns in the incoming acoustic signal and so it will not be discussed here. Breakdowns in the language system impair integration of the incoming acoustic signal into a language context and can produce faulty perception as well. Hearing health care professionals, who justifiably focus on the integrity of the incoming acoustic signal, often minimize the possibility that, even in the presence of a complete acoustic signal, faulty speech perception can occur. Under these circumstances, repair of a faulty incoming acoustic signal alone may not improve speech perception. The purpose of the next section is not to review the research in detail or debate theoretical issues concerning the role language plays in speech perception. The intent is to provide a sense of the variables, besides the incoming acoustic signal, that can impair classroom speech perception.

## LEARNING STRATEGIES

Returning to the model in Figure 2–1, speech perception is dependent upon receipt of specific acoustic information, and also upon the ability to relate these often ambiguous cues within the context of a language structure and use of circumstance. The cognitive activities that tie the acoustic signal to language are referred to as learning strategies (McCormick & Schiefelbusch, 1984). Breakdowns in the processes underlying the cognitive activities will impair the tie between the acoustic signal and language and so produce faulty perception. In the model shown in Figure 2–1, cognitive activities are represented by the boxes labeled attention, speech and language patterns, serial processing, and parallel processing.

### Attention

The ability to focus selectively on relevant features of an incoming acoustic signal is developmental. As the overall speech perceptual system matures, attention becomes more and more selective, and preferences or biases are established (Gibson, 1969). In the classroom, the ability to sustain focus is also very important. McCormick and Schiefelbusch (1984) refer to sustained attention as attention span or "on task" behavior. Educators relate the amount of time that a student is "on task" to success in learning (Cangelosi, 1991). Students who are unable to attend selectively to a signal or who cannot maintain sustained attention have a roadblock in the speech perceptual model shown in Figure 2–1. In the case of unselective attention, irrelevant signals will be as equally accessible to the speech perceptual process as are relevant signals. Noise and distractions may be focused on, in place of important speech signals. One need only think of the unselective attention of the autistic child to demonstrate the outcome of this kind of cognitive inability. The autistic child who delights in the hum of a refrigerator motor and totally disregards relevant commands is known only too well by the parents of such a child (Lovaas & Schreibman, 1979). It is hard to imagine how relevant analysis of speech can occur under these circumstances. The inability to sustain attention or stay "on task" can also block perception. Feagans, Sanyal, Henderson, Collier, and Applebaum (1987) hypothesized that the long-term effect of reduced hearing might well be that the student learns to become inattentive to language. Analysis of the incoming acoustic speech signal requires sustained attention, especially if the serial processing mode is in use. It is an irony that the student with the weakest ability to analyze language might well require the greatest attentional ability. Fortunately, it appears as if selective attention and sustained attention are skills that can be trained, if they are recognized early (Gibson, 1969; Feagans et al., 1987).

## Concept Formation

The details of concept formation in the language learning process are well beyond the scope of this chapter. It should be clear from the model in Figure 2–1, however, that a great deal of mental activity is involved in establishing speech and language patterns, serial and parallel processing, and the establishment of relationships between linguistic and non-linguistic signals and meaning. The reader is referred to Bloom and Lahey (1978) for a full discussion of the establishment of linguistic and non-linguistic concepts. A breakdown in concept formation affects perception because the manner and efficiency with which sensory information is organized determines to a large extent how useful it will be for future reference (McCormick & Schiefelbusch, 1984).

## Storage and Retrieval

Speech perception is accomplished by combining incoming acoustic signals with stored knowledge about language and knowledge of the context or circumstance of the communicative message. Speech perception is the resultant of an interaction among incoming and stored information (Denes & Pinson, 1993). A detailed discussion of the mechanisms that have been theorized for storage and retrieval of information in the nervous system are outside of the scope of this chapter. The reader is referred to the following references for more information: Geis and Hall (1977), Kobashigaw (1976), and McCormick and Schiefelbusch (1984). Whatever the memory process, incoming information must be organized in a manner that allows efficient storage, and in a manner that may be quickly retrieved. The retrieved information must be in a form that can be combined with incoming information and perceptions derived. Needless to say, considerable cognitive ability is required to accomplish these tasks in real time. Breakdowns in the storage and/or retrieval process can have devastating effects upon speech perception. One need only consider the communicative plight of the receptive or expressive aphasic person to understand the consequences in the extreme. Less severe breakdowns can slow the speech processing system down and impair the listener's ability to store and retrieve information in real time. Becoming lost in the incoming information stream most assuredly impairs perception.

The intent of this section was to provide a sense of the variables, besides the incoming acoustic signal, that can impair classroom speech perception. Selective and sustained attention abilities, competency in concept formation, and adequacy of storage and retrieval of information will not often be directly measured in children in the classroom. Most of these abilities can be inferred from the competency in the use of language exhibited by the children. Children showing signs of language problems are in need of clinical intervention, in order to maximize the use of the acoustic signal. If language problems are ignored, efforts to improve the incoming acoustic signal may not produce the expected improvements in classroom understanding. Authors have provided descriptions of behaviors that are characteristic of students with speech perceptual breakdowns (Cole & Wood, 1978; Owens, 1991). A modified version of the Cole and Wood checklist is shown in Table 2–1. While it is unlikely that a child with a subtle language disorder will exhibit all of these behaviors, any of the behaviors could signal a breakdown in the speech perceptual system and impair classroom understanding. Once identified, language problems must be managed by a speech-language pathologist as part of the overall strategy to improve classroom understanding.

**Table 2–1.** Modified version of the Cole and Wood checklist.

I.  **Auditory Reception**

    A.  Student disregards speech or all sounds

    ALWAYS (4)_____        SOMETIMES(2)_____        NEVER(0)_____

    B.  Student shows better responses in quiet than in noise

    ALWAYS (4)_____        SOMETIMES(2)_____        NEVER(0)_____

    C.  Student is overly sensitive to sound

    ALWAYS (4)_____        SOMETIMES(2)_____        NEVER(0)_____

    D.  Student relies on imitation of others speech

    ALWAYS (4)_____        SOMETIMES(2)_____        NEVER(0)_____

    E.  Student has difficulty following verbal instructions unless accompanied
        by visual demonstrations

    ALWAYS (4)_____        SOMETIMES(2)_____        NEVER(0)_____

    F.  Student has difficulty learning in a group situation

    ALWAYS (4)_____        SOMETIMES(2)_____        NEVER(0)_____

    G.  Student fails to remember what people say

    ALWAYS (4)_____        SOMETIMES(2)_____        NEVER(0)_____

    H.  Student does not generalize information from one experience to another

    ALWAYS (4)_____        SOMETIMES(2)_____        NEVER(0)_____

**Table 2–1.** *continued*

## II. Verbal Expression

A. Student produces a low quantity of verbalization

ALWAYS (4)_____ SOMETIMES(2)_____ NEVER(0)_____

B. Student uses an inadequate vocabulary

ALWAYS (4)_____ SOMETIMES(2)_____ NEVER(0)_____

C. Student demonstrates defective language structure

ALWAYS (4)_____ SOMETIMES(2)_____ NEVER(0)_____

D. Student cannot verbalize experience using sequential utterances

ALWAYS (4)_____ SOMETIMES(2)_____ NEVER(0)_____

E. Student demonstrates incorrect pronunciation (articulation) of words

ALWAYS (4)_____ SOMETIMES(2)_____ NEVER(0)_____

F. Student uses disorganized content within or among utterances

ALWAYS (4)_____ SOMETIMES(2)_____ NEVER(0)_____

G. Student is dependent upon gestures to express information

ALWAYS (4)_____ SOMETIMES(2)_____ NEVER(0)_____

H. Student is unusually literal in expressing ideas

ALWAYS (4)_____ SOMETIMES(2)_____ NEVER(0)_____

## III. Social-Emotional

A. Student has problems attending to pertinent tasks

ALWAYS (4)_____ SOMETIMES(2)_____ NEVER(0)_____

B. Student demonstrates an inability to inhibit behavior

ALWAYS (4)_____ SOMETIMES(2)_____ NEVER(0)_____

C. Student does not cope with change easily

ALWAYS (4)_____ SOMETIMES(2)_____ NEVER(0)_____

*Continued*

**Table 2–1.** *continued*

D. Student demonstrates a disorientation in time and space

ALWAYS (4)_____    SOMETIMES(2)_____            NEVER(0)_____

E. Student has immature self-help skills

ALWAYS (4)_____    SOMETIMES(2)_____            NEVER(0)_____

F. Student demonstrates preseverative behavior

ALWAYS (4)_____    SOMETIMES(2)_____            NEVER(0)_____

G. Student is hyperactive

ALWAYS (4)_____    SOMETIMES(2)_____            NEVER(0)_____

H. Student shows inappropriate emotional reactions

ALWAYS (4)_____    SOMETIMES(2)_____            NEVER(0)_____

I. Student is socially isolated

ALWAYS (4)_____    SOMETIMES(2)_____            NEVER(0)_____

J. Student is overly aggressive

ALWAYS (4)_____    SOMETIMES(2)_____            NEVER(0)_____

K. Student has limited interpersonal relationships

ALWAYS (4)_____    SOMETIMES(2)_____            NEVER(0)_____

## IV. Academic

A. Student has difficulty following verbal instructions or learning from verbal explanations

ALWAYS (4)_____    SOMETIMES(2)_____            NEVER(0)_____

B. Student has difficulty learning phonics

ALWAYS (4)_____    SOMETIMES(2)_____            NEVER(0)_____

C. Student demonstrates inadequate reading or spelling

ALWAYS (4)_____    SOMETIMES(2)_____            NEVER(0)_____

D. Student has poor comprehension of what is read

ALWAYS (4)_____    SOMETIMES(2)_____            NEVER(0)_____

**Table 2–1.** *continued*

E. Student produces writing with disorganized content

ALWAYS (4)_____   SOMETIMES(2)_____        NEVER(0)_____

F. Student has poor sentence structure in written work

ALWAYS (4)_____   SOMETIMES(2)_____        NEVER(0)_____

G. There is a discrepancy between student's academic achievement and potential

ALWAYS (4)_____   SOMETIMES(2)_____        NEVER(0)_____

|  | First Administration | Second Administration | Difference |
|---|---|---|---|
| Auditory Reception Score | _____ | _____ | _____ |
| Verbal Expression Score | _____ | _____ | _____ |
| Social-Emotional Score | _____ | _____ | _____ |
| Academic Score | _____ | _____ | _____ |

Adapted from Cole, P., & Wood, L. (1978). Differential diagnosis. In F. Martin (Ed), *Pediatric audiology*. Englewood Cliffs, NJ: Prentice Hall.

# SUMMARY

As was pointed out in the beginning of this chapter, the primary goal of the classroom educational process is to share experiences, exchange ideas, and transmit knowledge. This is accomplished by the students not only because they are able to receive specific acoustic information, but because they are able to relate these often ambiguous cues within the context of a language structure and use circumstantial cues. Not all acoustic information is specific enough and not all student speech perceptual systems are complete enough to allow the educational process to occur uninterrupted. It behooves the professional to identify and improve acoustic unspecificity caused by the classroom environment. It is equally important to determine any speech perceptual breakdowns individual students may have who must interact with the acoustic signal in the classroom, and to intervene clinically where possible. Taken together, there is a very good chance that the classroom educational process can be strengthened and student academic achievement improved.

# REFERENCES

Bloom, L., & Lahey, M. (1978). *Language development and language disorders.* New York: John Wiley and Sons.

Boothroyd, A. (1984). *Hearing impairments in young children.* Englewood Cliffs, NJ: Prentice-Hall.

Cangelosi, J. (1991). *Evaluating classroom instruction.* New York: Longman.

Cole, P., & Wood, L. (1978). Differential diagnosis. In F. Martin (Ed)., *Pediatric audiology.* Englewood Cliffs, NJ: Prentice-Hall.

Denes, P.B., & Pinson, E.N. (1993). *The speech chain.* New York: W. H. Freeman and Company.

Feagans, L., Sanyal, M., Henderson, F., Collier, A., & Appelbaum, M. (1987). Relationship of middle ear disease in childhood to later narrative and attentional skills. *Journal of Pediatric Psychology, 12*(4), 581–594.

Geis, M., & Hall, D. (1977). Encoding and incidental memory in children. *Journal of Experimental Child Psychology, 22,* 58–66.

Gibson, E. (1969). *Principles of perceptual learning and development.* New York: Appleton-Century-Crofts.

Kobashigawa, A. (1976). Retrieval strategies in the development of memory. In R.V. Kail & J. W. Hagan (Eds.), *Memory in cognitive development.* Hillsdale, NJ: Lawrence Earlbaum Associates.

Lovaas, O., & Schreibman, L. (1979). Stimulus overselectivity in autism: A review of research. *Psychological Bulletin, 86,* 1236–1254.

McCormick, L., & Schiefelbusch, R. (1984). *Early language intervention.* Columbus, OH: Merill Publishing.

Owens, R. (1991). *Language disorders: A functional approach to assessment and intervention.* Columbus, OH: Merill Publishing.

Sanders, D. (1993). *Management of hearing handicap.* Englewood Cliffs, NJ: Prentice-Hall.

# CHAPTER
# 3

# SPEECH PERCEPTION IN THE CLASSROOM

*Carl Crandell*
*Joseph Smaldino*

For optimal learning to occur in the academic environment, the teacher's voice must be highly intelligible to all children. Unfortunately, as discussed in Chapter 1, the acoustical environment present in the vast majority of classrooms is not conducive to the accurate recognition of speech, nor to the psychoeducational and psychosocial development of children. This chapter will examine the acoustical variables that can deleteriously affect speech recognition in classroom environments. Specifically, the following areas will be addressed: (1) the level of the teacher's voice, (2) ambient noise in the room, (3) reverberation time of the enclosure, and (4) the distance from the teacher to the student.

## LEVEL OF THE TEACHER'S VOICE

The speaking level of a teacher's voice can vary depending upon the amount of breath available, vocal tract resonance control, and ability to project the speech signal. Of course, the level of the teacher's voice in relation to the overall ambient noise in the classroom determines the signal-to-noise (S/N) ratio (S/N ratio will be discussed in detail in a later section). Since an acceptable S/N ratio is believed to be in the +15 or +20 dB range, the level of the teacher's voice becomes a variable for improving classroom acoustics. When the classroom acoustics are evaluated (see Chapter 5), the level of the teacher's voice will be determined and a goal set to increase that level 10 dB using amplification. It would be possible to attain S/N ratios in excess of +15 dB if the teacher's voice level was in the +5 to +10 dB S/N ratio range before amplification. Some teachers possess voice levels that would make very favorable S/N ratios possible with 10 dB of amplification. A study by Otting and Smaldino (1992) investigated the potential of training teachers to increase their speaking voice levels behaviorally in the classroom. It was possible to train teachers to increase their voice level 3 to 5 dB, but continual training was necessary to sustain the level enhancement. **It was concluded that classroom amplification is the most consistent means to increase teachers voice levels over and above their normal speaking level**.

## AMBIENT NOISE IN THE CLASSROOM

In a classroom setting, speech is seldom delivered to a child without interference from background noise and reverberation. Simply stated, background noise refers to any undesired auditory interference that hinders what a child wants, or needs, to hear. Accordingly, ambient noise in the classroom could be classified as construction work or automobile traffic outside of the school building, sounds generated in adjacent rooms, noises caused by ventilation systems or faulty fluorescent lights, as well as children talking, laughing, or moving within the classroom itself.

### Sources of Noise in the Classroom

Ambient noise in the classroom setting can originate from a number of possible sources (John & Thomas, 1957):

1. *External Noise Sources:* refers to noise that is generated from outside the school building, such as construction, automobile or aircraft traffic, and playground areas.
2. *Internal Noise Sources:* originate from within the school building, but outside the actual classroom. Classrooms adjacent to the cafeteria, gymnasium, and/or busy hallways often exhibit high internal noise levels.
3. *Classroom Noise Sources:* refers to noise that is generated within the classroom itself. Classroom noise includes chil-

dren talking or laughing, sliding of chairs or tables, shuffling of hard-soled shoes on non-carpeted floors, and school heating/cooling systems.

## Ambient Noise Levels in the Classroom

Because of the myriad of potential noise sources, classrooms often exhibit excessive levels of background noise. Sanders (1965), for example, measured the occupied and unoccupied noise levels of 47 classrooms in 15 different schools. The results from this investigation are presented in Table 3–1. Note that mean occupied noise levels ranged from 69 dB(B) in kindergarten classrooms to 52 dB(B) in classrooms utilized for children with hearing impairment (see Chapter 5 for a discussion of sound level meter weighting scales). Unoccupied classroom noise levels were approximately 10 dB lower than the occupied classroom settings, ranging from 58 dB(B) for kindergarten classrooms to 42 dB(B) in classrooms for the hearing impaired. Nober and Nober (1975) reported that the average intensity of 4 occupied elementary classrooms was 65 dB(A). Bess, Sinclair, and Riggs (1984) measured ambient noise levels in 19 classrooms for children with hearing impairment. Median unoccupied noise levels were 41 dB(A), 50 dB(B), and 58 dB(C). When the classroom was occupied with students, ambient noise levels increased to 56 dB(A), 60 dB(B), and 63 dB(C). Finally, in a recent investigation, Crandell and Smaldino (1994a) measured the ambient noise levels of 32 unoccupied classroom settings. Results from this investigation indicated classroom noise levels that were even higher than previously reported. Specifically, mean unoccupied noise levels were 51 dB(A) (range: 46–59 dB) and 67 dB(C) (range: 57–74 dB).

**Table 3–1.** Mean noise levels of various occupied classrooms.

| School Type | Noise Level (dBB) |
| --- | --- |
| Kindergarten | 69 |
| Elementary | 59 |
| High School | 62 |
| Hearing Impaired | 52 |

Adapted from Sanders, D. (1965). Noise conditions in normal school classrooms. *Exceptional Child, 31,* 344–353.

## Acoustical Criteria for Classroom Noise Levels

**The noise levels reported in the above-mentioned studies are distressing since acoustical recommendations for children and listeners with sensorineural hearing loss (SNHL) suggest that ambient noise levels in unoccupied classrooms should not exceed 30–35 dB(A) or a Noise Criteria Curve (NCC) of 20–25 dB for maximum speech recognition to occur** (see Chapter 5 for a discussion of NCCs). A review of the literature previously presented in this chapter, however, indicates that these acoustical recommendations are infrequently attained (Bess & McConnell, 1981; Crandell, 1991; Crandell & Bess, 1986; Finitzo-Hieber, 1988; McCroskey & Devens, 1975). McCroskey and Devens (1975) demonstrated that only one of nine elementary classrooms actually met these acoustical recommendations. Crandell and Smaldino (1994a) reported that none of 32 classrooms met recommended criteria (see Figure 3–1). In sum, it appears that ambient

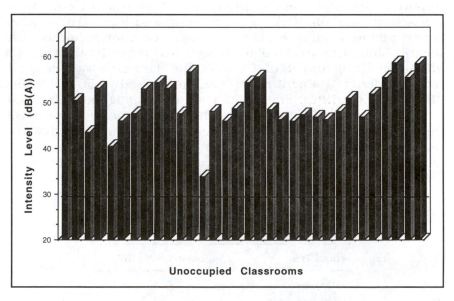

**Figure 3–1.** The average ambient noise levels, on the A-weighting scale, in 32 classrooms for the hearing impaired. The solid line indicates the recommended acoustical standard (noise level = 30 dB[A]). Figure used with permission from Crandell, C., & Smaldino, J. (1994). An update of classroom acoustics for children with hearing impairment. *The Volta Review,* (in press).

noise levels in the classroom are 10–15 dB higher than recommended standards. Moreover, and perhaps even more discouraging, the aforementioned studies suggest little improvement in classroom noise levels over the past several decades.

## Noise Effects on Speech Recognition

To appreciate the importance of excessive noise levels in the classroom, one must understand how noise affects a child's ability to understand speech. Ambient noise in the classroom affects a child's speech recognition by reducing, or *masking*, the highly redundant acoustic cues available in the teacher's voice. Masking refers to a phenomena in which the threshold of a signal, such as speech, is raised by the presence of another sound. For example, a listener may be able to understand a speaker who is speaking at normal conversational levels quite well in a quiet room. However, if the slide projector is turned on, the ventilation system begins to generate noise, and other persons begin to talk, the speaker will now have to raise the level of his or her voice for the listener to be able to understand what is being presented. If the speaker does not significantly raise the voice, then the noise in the classroom will mask much of the important acoustical information presented by the speaker. Consequently, the speech-recognition ability of the listener will be significantly reduced.

In general, the spectral energy of consonants is less intense than vowel energy. Because of these differences in spectral energy, noise in the classroom predominately reduces the recognition of consonant phonemes. Reductions in the recognition of consonants will significantly influence speech recognition as the vast majority of a listener's ability to understand speech is the result of consonantal energy.

The capability of noise to mask speech depends upon a number of acoustical parameters (Nabelek & Nabelek, 1985). These parameters include:

1. the long-term spectrum of the noise;
2. intensity fluctuations of the noise over time; and
3. the intensity of the noise relative to the intensity of speech.

In terms of the long-term spectrum of the noise, the most effective maskers for speech are those noises with a long-term spectra similar to the speech spectrum since they affect all speech frequencies to the same degree. Consequently, "classroom noise" tends to produce the greatest decreases in speech recognition, since the spectral content of the signal (teacher's voice) is spectrally similar to the spectra of the noise. Low-frequency noises in a classroom are often more effective maskers of speech than high-frequency noises because of a phenomenon called *upward spread of masking*. Because of upward spread of masking, noise tends to produce greater masking at frequencies below the frequency of a signal than at frequencies above the signal. For example, a 1000 Hz noise would more effectively mask a 2000 Hz signal than a 4000 Hz masker would. The fact that low-frequency noises have a greater effect on speech recognition than high-frequency noises is meaningful in the learning environment because the predominant long-term spectrum of noise found in classroom environments is frequently low frequency in nature.

Classroom noises which are continuous in nature are generally more effective maskers than interrupted or impulse noises. These differences in masking occur because continuous noises more effectively reduce the spectral-temporal information available in the speech signal. Continuous noises in the classroom include the hum of air conditioning or heating systems, faulty fluorescent lighting, and the long-term spectra of children talking.

In most learning environments, however, the fundamental consideration for speech recognition is the relationship among the intensity of the signal and the intensity of the ambient noise at the child's ear. This relationship is referred to as the *signal-to-noise ratio (S/N ratio), or message-to-competition ratio (MCR)*. To illustrate, if a speech signal is presented at 70 dB, and a noise is 65 dB, the S/N ratio (or MCR) is +5 dB. Generally speaking, speech-recognition ability is greatest at favorable S/N ratios and decreases as a function of reduction in S/N ratio (Cooper & Cutts, 1971; Crum, 1974; Finitzo-Hieber & Tillman, 1978; Miller, 1974; Nabelek & Pickett, 1974a, b). The relationship between S/N ratio and speech recognition is illustrated in Table 3–2. These data, taken from Crum (1974), show the word recognition scores of adult normal hearers at various S/N ratios (+12 dB, +6 dB, and 0 dB). As can be noted, recognition scores reach a plateau at high S/N ratios (95% at a S/N ratio of +12 dB), but are reduced as the S/N ratio becomes less favorable (81% and 46% at S/N ratios of +6 dB and 0 dB, respectively).

## Signal-to-Noise Ratios in the Classroom

Because of the excessive noise levels found in many learning environments, unfavorable S/N ratios have often been reported in the classroom setting. Specifically, as can be noted in Table 3–3, the range of S/N ratios for classrooms has been reported to be from approximately +5 dB to –7 dB (e.g., Blair, 1977; Finitzo-Hieber, 1988; Markides, 1986; Paul, 1967; Sanders, 1965).

## Criteria for Classroom Signal-to-Noise Ratios

Numerous acoustical and linguistical factors may influence appropriate classroom S/N ratios for children. For example, the type of acoustical stimuli that is presented (i.e., sentences, words, nonsense syllables, vowels, consonants) can affect perceptual ability in noise.

**Table 3–2.** Mean speech-recognition scores, in percent correct, of adults with normal hearing across various signal-to-noise ratios.

| Signal-to-Noise Ratio | Word Recognition |
|:---:|:---:|
| QUIET | 99.7 |
| +12 dB | 95.3 |
| +6 dB | 80.7 |
| 0 dB | 46.0 |

Adapted from Finitzo-Hieber, T., & Tillman, T. (1978). Room acoustics effects on monosyllabic word discrimination ability for normal and hearing-impaired children. *Journal of Speech and Hearing Research, 21*, 440–458.

**Table 3–3.** A summary of studies that have examined classroom signal-to-noise ratios.

| Investigation | Classroom S/N Ratio |
|:---|:---:|
| Sanders (1965) | +1 to +5 |
| Paul (1967) | +3 |
| Blair (1977) | –7 to 0 |
| Finitzo-Hieber (1988) | +1 to +4 |

Generally speaking, sentences contain considerably more linguistic information than do monosyllabic words, thus greater S/N ratios are required to perceive words accurately in noise. Moreover, as discussed previously, the power of the speaker's voice, the long-term spectrum of the noise, and fluctuations in the noise over time can all influence speech recognition. In general, speech recognition in adult listeners with normal hearing is not significantly affected until the speech and noise at are equal intensities (i.e., S/N ratio = 0 dB). Children and listeners with sensorineural hearing loss, however, require a more favorable S/N ratio to obtain adequate communicative efficiency in noise. **Overall, it has been recommended that S/N ratios in learning environments for these listeners should exceed at least +15 dB for maximum speech recognition** (Beranek, 1954; Borrild, 1978; Crandell, 1991, 1992; Fourcin et al., 1980; Gengel, 1971; Finitzo-Hieber, 1988; Finitzo-Hieber & Tillman, 1978; Niemoeller, 1968; Olsen, 1988). This recommendation is based on the findings that the speech recognition of children and listeners with SNHL tend to remain relatively constant at S/N ratios in excess of +15 dB, but deteriorate at poorer S/N ratios. Moreover, listening effort in children, particularly those children with hearing impairment, is minimum at S/N ratios exceeding +10 to +15 dB. To be able to provide this recommendation in most classroom settings, unoccupied noise levels must not exceed recommended criteria (30–35 dB(A) or NCC = 20–25 dB curve) (Borrild, 1978; Crandell, 1992; Fourcin et al., 1980; Gengel, 1971; Finitzo-Hieber, 1988; Finitzo-Hieber & Tillman, 1978; Neimoller, 1968; Olsen, 1988). Unfortunately, it is uncommon for appropriate S/N ratios to occur in the classroom. In fact, note that from Table 3–3 that many of the S/N ratios reported in typical listening environments were not even appropriate for *adult listeners* with normal hearing.

## Noise Effects on Psychoeducational/ Psychosocial Development

Prior to completing our discussion on noise, it is important to note that in addition to reductions in speech recognition, classroom noise can also compromise psychoeducational and psychosocial achievement in children (Ando & Nakane, 1975; Crook & Langdon, 1974; Dixon, 1976; Green, Pasternak, & Shore, 1982; Ko, 1979; Koszarny, 1978; Lehman & Gratiot, 1983; McCroskey & Devens, 1975; Sargent, Gidman, Humphreys, & Utley, 1980). Stated otherwise, noise has been demonstrated to affect academic performance, reading and spelling skills, concentration, attention, and student behavior adversely. Lehman and Gratiot (1983) reported that reductions in classroom noise (via acoustical modification) had a significant effect of increasing concentration, attention, and participatory behavior in children. Interestingly, the noise levels were reduced from typically reported noise levels of 35 to 45 dB(A) to the acoustical standard of 30 dB(A). Koszarny (1981) found that noise levels tend to more seriously affect concentration and attention in children with lower IQs and/or high anxiety levels. Green, Pasternak, and Shore (1982) found that classroom noise alone accounted for approximately 50–75% of the variance in reading delays of one year or more in elementary school children. Thus, classroom noise can affect not only the transmission of acoustical information, but also the operation of learning itself.

## Noise Effects on Teacher Performance

Classroom noise has also been shown to affect teacher performance (Crook & Langdon, 1974; Ko, 1979; Sargent, et al., 1980). For example, Ko (1979) obtained information from more than 1,200 teachers concerning the effects of noise in the classroom. Results indicated that noise related to classroom activities and traffic and/or airplane noise was related to teacher fatigue, increased tension and discomfort, and an interference with teaching and speech recognition. Additional studies (Crook & Langdon, 1974; Sargent et al., 1980) have reported that teachers exhibit a significantly higher incidence of vocal problems than does the general population. It is reasonable to assume that these vocal difficulties are caused, at least in part, by having to increase vocal output over the classroom noise during the school day.

## REVERBERATION IN THE CLASSROOM

A second acoustical variable that can be detrimental to speech recognition in the classroom is reverberation. Reverberation refers to the persistence or prolongation of sound within an enclosure as sound waves reflect off hard surfaces in a room (Kurtovic, 1975; Lochner & Burger, 1964; Nabelek & Pickett, 1974a,b). This prolongation of sound is usually considered the most important acoustical consideration that defines the acoustical climate of a classroom. Reverberation time (RT) refers to the amount of time it takes for a steady-state sound to decay 60 dB from its initial offset. For example, if a 100 dB SPL signal is delivered into a room and takes 2 seconds to be reduced to 40 dB SPL, the RT of that environment would be 2 seconds. In general, RT increases linearly with room volume and is inversely related to the amount of sound absorption in an environment. RT can be simply stated via the following formula:

$$RT = 0.049V/a$$

where *0.049* is a constant; *V* = volume of the enclosure; and *a* = the absorption characteristics in the room.

Based on this formula, it can be seen that larger classrooms tend to exhibit higher reverberation times than do classrooms having smaller dimensions. In addition, classrooms with irregular shapes, such as oblong, often exhibit higher reverberation times than do classrooms with more traditional quadrilateral dimensions. The presence, or absence, of absorptive surfaces within a classroom can also affect the reverberation time. Materials in the classroom with soft, rough-surfaced, and/or porous surfaces, such as cloth, fiberglass, and corkboard tend to be good absorbers of sound. Conversely, hard, smooth surfaces such as concrete, cinder block, and hard plaster are poor absorbers. Thus classrooms that contain bare cement walls, floors, and/or ceilings exhibit higher reverberation times than do classrooms which contain absorptive surfaces such as carpeting, draperies, and acoustical ceiling tile.

One index that is helpful in determining the reverberant characteristics of a classroom is the *absorption coefficient*. The absorption coefficient refers to the ratio of unreflected energy to incident energy present in a room. Table 3–4 presents average absorption coefficients of common room surfaces. Room surfaces with an absorption coefficient of 1.00 would technically absorb 100 percent of all refections, while a surface structure with an absorption coefficient of 0.00 would reflect all of the incident sound. Note that the absorption coefficients are frequency dependent. Specifically, most surface materials in a classroom do not absorb low-frequency sounds as effectively as higher frequencies. Because of these absorption characteristics, classroom reverberation is often shorter at higher frequencies than in lower-frequency regions. Generally, surfaces are not considered as being "absorptive" until they reach an absorption coefficient of 0.20.

## Reverberation Effects on Speech Recognition

Reverberation in the classroom influences the recognition of speech through the "smearing" or masking of direct sound energy

**Table 3–4.** Sound absorption coefficient for various types of acoustical treatments and for building materials.

| Material | Frequency (Hz) | | | | | |
|---|---|---|---|---|---|---|
| | 125 | 250 | 500 | 1k | 2k | 4k |
| **Walls** | | | | | | |
| Brick | 0.03 | 0.03 | 0.03 | 0.04 | 0.05 | 0.07 |
| Concrete painted | 0.10 | 0.05 | 0.06 | 0.07 | 0.09 | 0.08 |
| Window glass | 0.35 | 0.25 | 0.18 | 0.12 | 0.07 | 0.04 |
| Marble | 0.01 | 0.01 | 0.01 | 0.02 | 0.02 | 0.00 |
| Plaster or concrete | 0.12 | 0.09 | 0.07 | 0.05 | 0.05 | 0.04 |
| Plywood | 0.28 | 0.22 | 0.17 | 0.09 | 0.10 | 0.11 |
| Concrete block, coarse | 0.36 | 0.44 | 0.31 | 0.29 | 0.39 | 0.25 |
| Heavyweight drapery | 0.14 | 0.35 | 0.55 | 0.72 | 0.70 | 0.65 |
| Fiberglass wall treatment, 1 inch (2.5 cm) | 0.08 | 0.32 | 0.99 | 0.76 | 0.34 | 0.12 |
| Fiberglass wall treatment, 7 inch (17.8 cm) | 0.86 | 0.99 | 0.99 | 0.99 | 0.99 | 0.99 |
| Wood paneling on glass fiber blanket | 0.40 | 0.99 | 0.80 | 0.50 | 0.40 | 0.30 |
| **Floors** | | | | | | |
| Wood parquet on concrete | 0.04 | 0.04 | 0.07 | 0.60 | 0.06 | 0.07 |
| Linoleum | 0.02 | 0.03 | 0.03 | 0.03 | 0.03 | 0.02 |
| Carpet on concrete | 0.02 | 0.06 | 0.14 | 0.37 | 0.60 | 0.65 |
| Carpet on foam rubber padding | 0.08 | 0.24 | 0.57 | 0.69 | 0.71 | 0.73 |
| **Ceilings** | | | | | | |
| Plaster, gypsum, or lime on lath | 0.14 | 0.10 | 0.06 | 0.05 | 0.04 | 0.03 |
| Acoustic tiles 2/3 inch (1.6 cm), suspended 16 inches (40.3 cm) | 0.25 | 0.28 | 0.46 | 0.71 | 0.86 | 0.93 |
| Acoustic tiles 1/2 inch (1.2 cm), suspended 16 inches (40.6 cm) from ceiling | 0.52 | 0.37 | 0.50 | 0.69 | 0.79 | 0.78 |
| The same as above, but cemented directly to ceiling | 0.10 | 0.22 | 0.61 | 0.56 | 0.74 | 0.72 |
| High absorptive panels, 1 inch (2.5 cm), suspended 16 inches (40.6 cm) from ceiling | 0.58 | 0.88 | 0.75 | 0.99 | 1.00 | 0.96 |
| **Others** | | | | | | |
| Upholstered seats | 0.19 | 0.37 | 0.56 | 0.67 | 0.61 | 0.59 |
| Audience in upholstered seats | 0.39 | 0.57 | 0.80 | 0.94 | 0.92 | 0.87 |
| Grass | 0.11 | 0.26 | 0.60 | 0.69 | 0.92 | 0.99 |
| Soil | 0.15 | 0.25 | 0.40 | 0.55 | 0.60 | 0.60 |
| Water surface | 0.01 | 0.01 | 0.01 | 0.02 | 0.02 | 0.03 |

Adapted from Lipscomb, D. (1978). *Noise in audiology.* Austin, TX: Pro-Ed.

by reflected energy (Kurtovic, 1975; Nabelek & Pickett, 1974a, b). To explain, in a reverberant environment, the reflected signals reaching the ear are temporally delayed and overlap with the direct signal, resulting in degradation of the speech signal (Houtgast, 1981; Kurtovic, 1975; Lochner & Burger, 1964; Nabelek & Pickett, 1974a, b). In general, speech-recognition scores decline with increases in RT (e.g., Finitzo-Hieber & Tillman, 1978; Gelfand & Silman, 1979; Neuman & Hochberg, 1983; Nabelek & Pickett, 1974a, b). This trend, taken from the seminal article by Finitzo-Hieber and Tillman (1978), is illustrated in Table 3–5. Note that while children with normal hearing obtained mean word recognition scores of 92.5 percent at a reverberation time of 0.4 second, recognition ability decreased to 76.5 percent at a reverberation time of 1.2 seconds.

The speech recognition of children and listeners with SNHL is more adversely affected by elevations in reverberation than it is in adult listeners with normal hearing (see Table 3–5) (Finitzo-Hieber & Tillman, 1978; Neuman & Hochberg, 1983; Nabelek & Pickett, 1974a, b; Nabelek & Robinson, 1982). For example, speech recognition in adult normal hearers is not significantly reduced until reverberation times exceed approximately one second (Crum, 1974, Gelfand & Silman, 1979; Nabelek & Pickett, 1974a, b). Listeners with SNHL and children, however, need reverberation times no greater than 0.4 second for optimum communicative efficiency (Crandell & Bess, 1986; Crandell, 1991, 1992; Finitzo-Hieber, 1988; Finitzo-Hieber & Tillman, 1978; Neimoller, 1968; Olsen, 1988).

**Table 3–5.** Mean speech-recognition scores, in percent correct, of children with normal hearing and hearing impairment (with/without hearing aids) for monosyllabic words across various reverberation times.

| Reverberation Time (Seconds) | GROUPS | | |
| :---: | :---: | :---: | :---: |
| | Normal | Hearing Impaired | Hearing Impaired (Aided) |
| 0.0 | 94.5 | 87.5 | 83.0 |
| 0.4 | 82.8 | 69.0 | 74.0 |
| 1.2 | 76.5 | 61.8 | 45.0 |

Adapted from Finitzo-Hieber, T., & Tillman, T. (1978). Room acoustics effects on monosyllabic word discrimination ability for normal and hearing-impaired children. *Journal of Speech and Hearing Research, 21,* 440–458.

## Reverberation Times in Classrooms

The reverberation times of classrooms are often too high for optimum speech recognition to occur for children. Specifically, as can be noted from Table 3–6, the range of reverberation for unoccupied classroom settings is typically reported to be from 0.4 to 1.2 seconds (Crandell, 1991; Crandell & Smaldino, 1994; Finitzo-Hieber, 1988; Kodaras, 1960; McCroskey & Devens, 1975; Nabelek & Pickett, 1974a, Ross, 1978).

## Recommended Criteria for Reverberation Times in Classroom

**The levels of classroom reverberation presented in Table 3–6 are cause for concern since acoustical standards recommend that reverberation time should not exceed 0.4 seconds for maximum speech recognition in the classroom** (Bradley, 1986; Crandell & Bess, 1986; Crum, 1974; Finitzo-Hieber & Tillman, 1978; Neimoller, 1968; Olsen, 1977, 1981). It is obvious from this table that appropriate levels of reverberation seldom occur in the classroom setting. Crandell and Smaldino (1994a), for example, reported that only 9 of 32 classrooms (28%) exhibited reverberation times less than 0.4 second (see Figure 3–2). In sum, classroom reverberation times are approximately 0.1 to 0.8 second longer than recommended. As with ambient noise levels, classroom reverberation levels have also shown minimal change over the past several decades.

## CLASSROOM NOISE AND REVERBERATION

Thus far, this chapter has addressed the individual effects of noise and reverberation on speech recognition. However, in the classroom setting, these acoustic phenomena do not occur in iso-

**Table 3–6.** A summary of studies that have examined classroom reverberation times.

| Investigation | Classroom Reverberation |
| --- | --- |
| Kodaras (1960) | 0.4 to 1.1 |
| McCroskey & Devens (1975) | 0.6 to 1.0 |
| Nabelek & Picket (1974a) | 0.5 to 1.0 |
| Crandell & Smaldino (1994) | 0.35 to 1.2 |

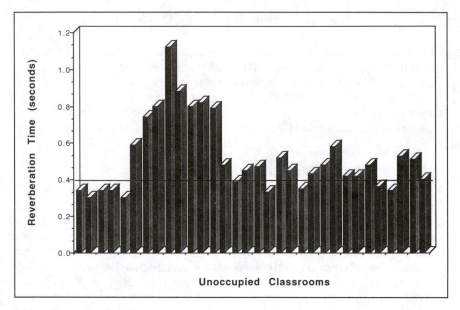

**Figure 3–2.** The average reverberation time (500, 1000, and 2000 Hz) in 32 classrooms for the hearing impaired. The solid line indicates the recommended acoustical standard (reverberation time = 0.4 second). Figure used with permission from Crandell, C., & Smaldino, J. (1994). An update of classroom acoustics for children with hearing impairment. *The Volta Review*, (in press).

lation. Therefore, a discussion of the interaction of noise and reverberation is necessary. In the classroom, noise and reverberation combine synergistically to affect speech recognition (Crandell & Bess, 1986; Crum, 1974; Finitzo-Hieber & Tillman, 1978; Nabelek & Pickett, 1974a, b). That is, the interaction of noise and reverberation adversely affects speech recognition to a greater extent than the sum of both effects taken independently. To illustrate, if a listener experiences a reduction in speech recognition of 5 percent in a noisy listening environment and a reduction of 5 percent in a reverberant setting, recognition deficits may actually equate to 20–30 percent in a listening environment that contains both noise and reverberation. These synergistic effects appear to occur because when noise and reverberation are combined, reflections fill in the temporal gaps in the noise, making it more steady-state in nature (recall from the **Speech Recognition in Noise** section that the most effective maskers of speech are those noises with a spectrum similar to the speech spectrum). As with noise and reverberation in isolation, research indicates that children and individuals with SNHL experience greater speech-recognition difficulties in noise *and* reverberation than adult normal hearers (Crandell & Bess, 1986; Crum, 1974; Finitzo-Hieber & Tillman, 1978; Nabelek & Nabelek, 1985).

## SPEAKER-TO-LISTENER DISTANCE

A final acoustical factor that influences speech recognition in the classroom is the distance from the teacher to the student. Distribution of sound in a classroom, as a function of speaker-listener distance, is illustrated in Figure 3–3. In this figure, *Line a* represents the sound pressure level (SPL) of the direct sound in the room. *Curve b* depicts the total sound pressure in the room. The total sound pressure is the sum of the reflected and direct sound energy in the classroom, and is often called the reverberant, or indirect sound field. *Point Dc* (indicated

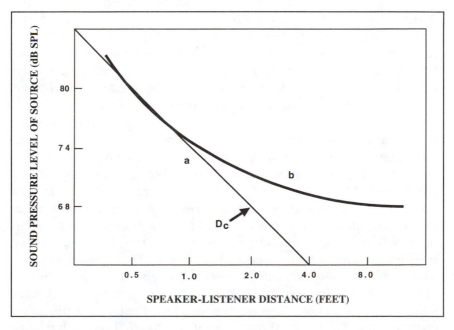

**Figure 3–3.** The distribution of sound in a classroom, as a function of speaker-listener distance. Adapted from Nabelek, A., & Nabelek, I. (1985). Room acoustics and speech perception. In J. Katz (Ed.), *Handbook of clinical audiology* (3rd ed). Baltimore: Williams & Wilkins.

by the arrow) indicates the critical distance of the room. The critical distance refers to that point in a room in which the intensity of the direct sound is equal to the intensity of the reverberant sound.

At distances relatively close to the child, the direct sound field predominates in the listening environment. In this sound field, sound waves are propagated outward in a spherical pattern and are transmitted from the teacher to the child with minimal interference from room surfaces. Direct sound pressure follows the principle of the *inverse square law*. This law states that sound level decreases 6 dB for every doubling of distance from the sound source. Because of the inverse square law, the direct sound field in a classroom is dominant only at distances close to the teacher.

As the child moves away from the teacher, the indirect or reverberant field now dominates the listening environment. The indirect sound field originates at the critical distance of the room. Operationally, critical distance can be defined by the following formula:

$$Dc = (0.20)(VQ/nRT)^{-1/2}$$

where $V$ = volume of the room in cubic meters, $Q$ = directivity factor of the source (the human voice is approximately 2.5), $n$ = number of sources, and $RT$ = reverberation time of the enclosure at 1400 Hz. In a average-sized classroom (150 $m^3$) with a typical reverberation time of 0.6 second, the critical distance would be slightly greater than 9 feet or 3 meters from the teacher. Thus, many, if not most, children in classrooms will be in the indirect sound field.

Beyond the critical distance, the direct sound from the speaker arrives at the listener initially, but is followed by reverberated signals that are composed of the original wave which has now been reflected off the ceiling, walls, and floor. Because there is a linear decrease in the intensity of the direct sound, and because the absorptive characteristics of structures in the room absorb some frequencies more than others, the reflected sound reaching the listener will contain a different acoustical content in the intensity, frequency, and temporal domains. Recall that these reflected waves can significantly affect speech recognition, particularly in children and listeners with SNHL. Interestingly, these reflected sound waves often result in an approximately equal distribution of energy beyond the critical distance of the room. This uniform distribution of sound energy, however, may not occur in larger listening environments, particularly if the power of the sound source is limited (Klein, 1971; Nabelek & Nabelek, 1985; Peutz, 1971).

## Effects of Distance on Speech Recognition in the Classroom

The distance a child is from the teacher can strongly influence speech recognition. Specifically, when the child is within the critical distance (the direct sound field), reverberation will have minimal effects on speech recognition. Beyond the critical distance (the indirect sound field), however, these reflections can significantly reduce speech recognition, particularly if there is sufficient spectral and/or intensity changes in the reflected sound to interfere with the recognition of the direct sound. Because of the effects of reverberation and reductions in S/N ratio, speech-recognition scores decrease until the critical distance of the room is reached (Crandell & Bess, 1986, Crandell, 1991; Klein, 1971; Peutz, 1971). Beyond the critical distance, recognition ability tends to remain essentially constant in the classroom. This finding suggests that speech-recognition ability can only be improved by decreasing the distance between a speaker and listener within the critical distance of the room. This finding also suggests that the recommendation of preferential seating has significant practical limitations. In typical classrooms the critical distance for maximum speech recognition is present only at distances relatively close to the teacher. Hence, the simple recommendation of preferential seating is often not enough to ensure an appropriate listening environment for many children.

## CONCLUSIONS

This chapter has addressed the effects of noise, reverberation, and distance on speech-recognition ability in the classroom. It was demonstrated that most learning environments demonstrate extremely poor acoustical conditions. Such findings are alarming as it is well recognized that inappropriate classroom acoustics can deleteriously affect not only speech perception, but also academic achievement.

## REFERENCES

Ando,Y., & Nakane, Y. (1975). Effects of aircraft noise on the mental work of pupils. *Journal of Sound and Vibration, 43*(4), 683–691.

Beranek, L. (1954). *Acoustics.* New York: McGraw-Hill.

Berg, F. (1993). *Acoustics and sound systems in schools.* San Diego: Singular Publishing Group.

Bess, F., & McConnell, F. (1981). *Audiology, education and the hearing-impaired child.* St. Louis: C.V. Mosby.

Bess, F., Sinclair, J., & Riggs, D. (1984). Group amplification in schools for the hearing-impaired. *Ear and Hearing, 5,* 138–144.

Blair, J. (1977). Effects of amplification, speechreading, and classroom environment on reception of speech. *Volta Review, 79,* 443–449.

Borrild, K. Classroom acoustics. In M. Ross & T. Giolas (Eds.), *Auditory management of hearing impaired children,* (pp. 145–179). Baltimore: University Park Press.

Bradley, J. (1986). Speech intelligibility studies in classrooms. *Journal of the Acoustical Society of America, 80*(3), 846–854.

Cooper, J., & Cutts, B. (1971). Speech discrimination in noise. *Journal of Speech and Hearing Research, 14,* 332–337.

Crandell, C. (1991). Classroom acoustics for normal-hearing children: Implications for rehabilitation. *Educational Audiology Monograph, 2*(1), 18–38.

Crandell, C. (1992). Classroom acoustics for hearing-impaired children. *Journal of the Acoustical Society of America, 92*(4), 2470.

Crandell, C., & Bess, F. (1986). Speech recognition of children in a "typical" classroom setting. *ASHA, 29,* 87.

Crandell, C., & Smaldino, J. (1992). Sound-Field Amplification in the Classroom. *American Journal of Audiology, 1*(4), 16–18.

Crandell, C., & Smaldino, J. (1994). An update of classroom acoustics for children with hearing impairment. *The Volta Review,* (in press).

Crandell, C., & Smaldino, J. (1994). The importance of room acoustics. In R. Tyler & D. Schum (Eds.), *Assistive listening devices.* Baltimore: William and Wilkins (in press).

Crook, M., & Langdon, F. (1974). The effects of aircraft noise in schools around London airport. *Journal of Sound and Vibration, 34,* 221–232.

Crum, D. (1974). *The effects of noise, reverberation, and speaker-to-listener distance on speech understanding.* Unpublished doctoral dissertation, Northwestern University, Evanston, IL.

Dixon, P. (1976). The effects of noise on childrens psychomotor, perceptual, and cognitive performance. Unpublished doctoral dissertation, University of Michigan.

Finitzo-Hieber, T. (1988). Classroom acoustics. In R. Roeser & M. Downs (Eds.), *Auditory disorders in school children* (2nd ed. pp. 221–233). New York: Thieme-Stratton.

Finitzo-Hieber, T., & Tillman, T. (1978). Room acoustics effects on monosyllabic word discrimination ability for normal and hearing-impaired children. *Journal of Speech and Hearing Research, 21,* 440–458.

Fourcin, A., Joy, D., Kennedy, M., Knight, J., Knowles, S., Knox, E., Martin, M., Mort, J., Penton, J., Poole, D., Powell, C., & Watson, T. (1980). Design of educational facilities for deaf children. *British Journal of Audiology,* Supplement #3.

Gelfand, S., & Silman, S. (1979). Effects of small room reverberation upon the recognition of some consonant features. *Journal of the Acoustical Society of America, 66*(1), 22–29.

Gengel, R. (1971). Acceptable signal-to-noise ratios for aided speech discrimination by the hearing impaired. *Journal of Auditory Research, 11*, 219–222.

Green, K., Pasternak, B., & Shore, B. (1982). Effects of aircraft noise on reading ability of schools age children. *Archives of Environmental Health, 37*, 24–31.

Houtgast, T. (1981). The effect of ambient noise on speech intelligibility in classrooms. *Applied Acoustics, 14*, 15–25.

John, J., & Thomas, H. (1957). Design and construction of schools for the deaf. In A. Ewing (Ed.), Educational guidance and the deaf child. Washington, DC: *Volta Review.*

Klein, W. (1971). Articulation loss of consonants as a criterion for speech transmission in a room. *Journal of the Audio Engineering Society, 19*, 920–922.

Ko, N. (1979). Response of teachers to aircraft noise. *Journal of Sound and Vibration, 62*, 277–292.

Kodaras, M. (1960). Reverberation times of typical elementary school settings. *Noise Control, 6*, 17–19.

Koszarny, Z. (1978). Effects of aircraft noise on the mental functions of school children. *Archives of Acoustics, 3*, 85–105.

Kurtovic, H. (1975). The influence of reflected sound upon speech intelligibility. *Acoustica, 33*, 32–39.

Leavitt, R., & Flexer, C. (1991). Speech degradation as measured by the Rapid Speech Transmission Index (RASTI). *Ear & Hearing, 12*, 115–118.

Lehman, A., & Gratiot, A. (1983). Effects du bruit sur les enfants a l'ecole. Proceedings of the 4th congress on noise as a public health problem (pp. 859–862). Milano: Centro Ricerche e Studi Amplifon.

Lochner, J., & Burger, J. (1964). The influence of reflections in auditorium acoustics. *Journal of Sound and Vibration, 4*, 426–454.

Markides, A. (1986). Speech levels and speech-to-noise ratios. *British Journal of Audiology, 20*, 115–120

McCroskey, F., & Devens, J. (1975). Acoustic characteristics of public school classrooms constructed between 1890 and 1960. *NOISEXPO Proceedings*, 101–103.

Miller, G. (1974). Effects of noise on people. *Journal of the Acoustical Society of America, 56*, 724–764.

Nabelek, A., & Nabelek, I. (1985). Room acoustics and speech perception. In J. Katz (Ed.), *Handbook of clinical audiology* (3rd ed). Baltimore:Williams & Wilkins.

Nabelek, A., & Pickett, J. (1974a). Monaural and binaural speech perception through hearing aids under noise and reverberation with normal and hearing-impaired listeners. *Journal of Speech and Hearing Research, 17*, 724–739.

Nabelek, A., & Pickett, J. (1974b). Reception of consonants in a classroom as affected by monaural and binaural listening, noise, reverberation, and hearing aids. *Journal of the Acoustical Society of America, 56*, 628–639.

Nabelek, A., & Robinson, P. (1982). Monaural and binaural speech perception in reverberation for listeners of various ages. *Journal of the Acoustical Society of America, 71*(5), 1242–1248.

Neuman, A., & Hochberg, I. (1983). Children's perception of speech in reverberation. *Journal of the Acoustical Society of America, 73*(6), 2145–2149.

Niemoller, A. (1968). Acoustical design of classrooms for the deaf. *American Annals of the Deaf, 113*, 1040–1045.

Nober, L., & Nober, E. (1975). Auditory discrimination of learning disabled children in quiet and classroom noise. *Journal of Learning Disabilities, 8*, 656–677

Olsen, W. (1977). Acoustics and amplification in classrooms for the hearing impaired. In F.H. Bess (Ed.), *Childhood deafness: causation, assessment and management.* New York: Grune & Stratton.

Olsen, W. (1981). The effects of noise and reverberation on speech intelligibility. In F.H. Bess, B.A. Freeman, & J.S. Sinclair (Eds.), *Amplification in education.* Washington, DC: Alexander Graham Bell Association for the Deaf.

Olsen, W. (1988). Classroom acoustics for hearing-impaired children. In F.H. Bess (Ed.), *Hearing Impairment in children.* Parkton, MD: York Press.

Paul, R. (1967). *An investigation of the effectiveness of hearing aid amplification in regular and special classrooms under instructional conditions.* Unpublished doctoral dissertation, Wayne State University.

Ottring, S., Smaldino, J., Plakke, B., & Bozik, M. (1992). *Comparison of two methods of improving classroom S/N ratio.* Paper presented at the American Speech, Language, and Hearing Association Convention, San Antonio, TX.

Peutz, V. (1971). Articulation loss of consonants as a criterion for speech transmission in a room. *Journal of the Audio Engineering Society, 19*, 915–919.

Ross, M. (1978). Classroom acoustics and speech intelligibility. In J. Katz (Ed.), *Handbook of clinical audiology.* Baltimore: Williams and Wilkins.

Sanders, D. (1965). Noise conditions in normal school classrooms. *Exceptional Child, 31*, 344–353.

Sargent, J., Gidman, M., Humphreys, M., & Utley, W. (1980). The disturbance caused by school teachers to noise. *Journal of Sound and Vibration, 62*, 277–292.

# CHAPTER
# 4

# SPEECH PERCEPTION IN SPECIFIC POPULATIONS

*Carl Crandell*
*Joseph Smaldino*
*Carol Flexer*

It is well recognized that listeners with sensorineural hearing loss (SNHL) experience greater difficulty understanding speech in noise and/or reverberation than do normal hearers. What is not as well recognized is that there are a number of populations of children with "normal hearing" sensitivity who also experience significant difficulties understanding noisy or reverberated speech. **As illustrated in Table 4–1, these listeners include children with fluctuating conductive hearing loss, learning disabilities, articulation disorders, central auditory processing deficits, language disorders, minimal degrees of sensorineural hearing loss (pure-tone sensitivity from 15 to 25 dB HL), unilateral hearing loss, developmental delays, and children for whom English is a second language.** At present, however, the auditory, linguistic, and/or cognitive mechanism(s) responsible for perceptual difficulties in these populations remain unclear. This chapter will examine the effects of commonly-reported classroom acoustics on the speech recognition of children with SNHL as well as children with "normal hearing." Moreover, this chapter will address the influence of sound-field amplification on the perceptual abilities and psychoeducational development of these populations.

## CHILDREN WITH SENSORINEURAL HEARING LOSS

Recent estimates suggest that more than 10 million school-age children exhibit some degree of SNHL. As discussed in Chapters 2

**Table 4–1.** Several populations of children with normal hearing sensitivity that experience greater speech-perception difficulties in degraded listening environments.

| "NORMAL HEARERS" |
| --- |
| YOUNG (<13–15 YEARS OLD) |
| CONDUCTIVE HEARING LOSSES |
| ARTICULATION DISORDERS |
| LANGUAGE DISORDERS |
| LEARNING DISABLED |
| NON-NATIVE ENGLISH |
| CENTRAL AUDITORY PROCESSING DEFICITS |
| MINIMAL SENSORINEURAL HEARING LOSSES |
| UNILATERAL HEARING LOSSES |

and 3, children with SNHL require considerably better acoustical environments than do normal hearers in order to process and understand speech. **Recall that children with SNHL require S/N ratios to surpass +15 dB and reverberation times no longer than 0.4 second for maximum communicative efficiency.** An illustration of the effects of hearing impairment on speech perception is presented in Table 4–2. These data, taken from Finitzo-Hieber and Tillman (1978), show the speech-recognition abilities of children (8 to 12 years of age) with mild-to-moderate degrees of SNHL (with and without amplification) compared to children with normal hearing sensitivity. Speech perception was assessed with monosyllabic words under various S/N ratios (quiet, +12, +6, 0) and reverberation times (T = 0.0, 0.4, and 1.2 seconds). Results from this investigation reveal several trends. First, these data indicate that the children with hearing impairment performed significantly poorer than did the children with normal hearing across most listening conditions. Second, the performance decrement between the two groups increased as the listening environment became less favorable. For example, in what would be an extremely good classroom environment (S/N ratio = +12 dB; T=0.4 second) children with hearing impairment obtained recognition scores of only 60 percent compared to 83 percent for the normal hearers (a 13 percent difference). In acoustical conditions more commonly reported in the classroom (S/N ratio = +6 dB; T=1.2 seconds), the performance difference increased to 27 percent, with the children having SNHL obtaining recognition scores of just 27 percent. Although not shown the addition of a hearing aid may not improve perceptual ability and, in fact, make understanding even more difficult in many listening conditions.

**Table 4–2.** Mean speech-recognition scores, in percent correct, of children with normal hearing and hearing impairment for monosyllabic words across various signal-to-noise ratios and reverberation times.

| Test Condition | Normal Hearing | Hearing Impaired |
|---|---|---|
| *RT = 0.0 Second* | | |
| Quiet | 94.5 | 83.0 |
| +12 dB | 89.2 | 70.0 |
| +6 dB | 79.7 | 59.5 |
| 0 dB | 60.2 | 39.0 |
| *RT = 0.4 Second* | | |
| Quiet | 92.5 | 74.0 |
| +12 dB | 82.8 | 60.2 |
| +6 dB | 71.3 | 52.2 |
| 0 dB | 47.7 | 27.8 |
| *RT = 1.2 Second* | | |
| Quiet | 76.5 | 45.0 |
| +12 dB | 68.8 | 41.2 |
| +6 dB | 54.2 | 27.0 |
| 0 dB | 29.7 | 11.2 |

Adapted from Finitzo-Hieber, T., & Tillman, T. (1978). Room acoustics effects on monosyllabic word discrimination ability for normal and hearing-impaired children. *Journal of Speech and Hearing Research, 21,* 440–458.

Few investigations have examined the effects of sound field amplification on the speech perception of children with SNHL (Crandell & Schmitt, 1994; Blair, Myrup, & Viehweg, 1989). Blair, Myrup, and Viehweg (1989) assessed word-recognition ability in a classroom under three different conditions of amplification: (1) sound-field amplification in conjunction with the child's personal hearing aid, (2) personal hearing aid, and (3) personal FM system with a neckloop configuration. Subjects consisted of 10 children with mild to moderate degrees of SNHL. As expected, the personal FM unit provided the greatest benefit in speech recognition (87 percent). However, the combination of the child's personal hearing aid and a sound-field amplification system produced a significant increase in speech perception over just the utilization of a personal hearing aid (82 percent compared to 70 percent).

Crandell and Schmitt (1994) examined the effects of sound field amplification on the psychoeducational and psychosocial achievement of children with SNHL. Subjects consisted of 13 children, in kindergarten to sixth grade, all of whom were utilizing personal amplification (hearing aids). The Screening Instrument for Targeting Educational Risk (SIFTER) evaluation, filled out by the student's teacher, was obtained before and after sound-field amplification use. Although post-amplification scores showed an improvement in all five areas of the SIFTER evaluation, only academic achievement scores proved to be statistically significant. Overall, these data suggest that sound-field amplification systems can offer, at least, minimal benefits to children utilizing personal amplification in the classroom. It seems reasonable to assume that sound field systems can provide the most benefit to such children while malfunctioning hearing aids or auditory trainers are being repaired.

## CHILDREN WITH MINIMAL SENSORINEURAL HEARING LOSS

Despite the importance of providing an appropriate acoustic environment for children with SNHL, there remains a paucity of data concerning the communicative efficiency of children with minimal, or "borderline," degrees of SNHL. That is, children with pure-tone thresholds between 15 and 30 dB HL through the speech-frequency range. Although the incidence of minimal hearing impairment in children is not known, it is well recognized that incidence rates increase as a function of decreasing hearing loss (Bess, 1985; Bess & McConnell, 1981; Crandell, 1993a; Davis, Elfenbein, Schum, & Bentler, 1986). Therefore, it is reasonable to assume that a significant number of school-age children may exhibit minimal degrees of SNHL.

At present, only two investigations have examined the effects of noise and/or reverberation on minimally hearing-impaired children. Boney and Bess (1984) demonstrated that children with minimal degrees of sensorineural hearing loss (pure-tone thresholds from 15 to 30 dB from 500 to 2000 Hz) experienced greater difficulty understanding speech degraded by noise and reverberation than did children with normal-hearing sensitivity. Specifically, word- and sentence-recognition scores were obtained in four listening conditions: (1) quiet, (2) reverberation (T = 0.8 second), (3) noise (S/N ratio = +6 dB), and (4) noise and reverberation (S/N ratio = +6 dB; T = 0.8 second). Word recognition results from this investigation (see Figure 4–1) indicated that the children with minimal hearing loss performed poorer than the control group, particularly in the degraded listening conditions.

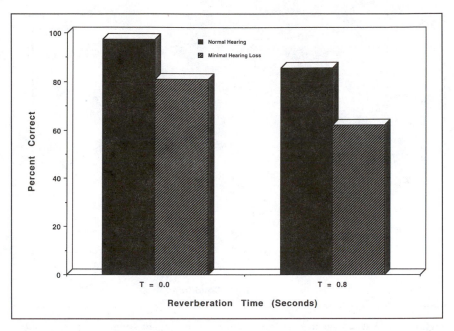

**Figure 4–1.** Speech perception of children with minimal degrees of sensorineural hearing loss (pure-tone thresholds from 15 to 30 dB from 500 to 2000 Hz) in a non-reverberant (T = 0.0 second) and reverberant (T = 0.8 second) listening condition. Adapted from Boney, S., & Bess, F. (1984). *Noise and reverberation effects in minimal bilateral sensorineural hearing loss.* Paper presented at The American Speech, Language, and Hearing Association Convention, San Francisco, CA.

Crandell (1993a) examined the speech perception of children with minimal degrees of sensorineural hearing loss at commonly-reported classroom S/N ratios of +6, +3, 0, –3, and –6 dB. The minimally hearing-impaired children exhibited pure-tone averages (0.5 kHz–2kHz) from 15 to 25 dB HL. Speech perception was assessed with the Bamford-Koval-Bench (BKB) Standard Sentence test. Multitalker babble from the Speech Perception in Noise (SPIN) test was used as the noise competition. Mean sentence recognition scores (in percent correct) as a function of S/N ratio are presented in Figure 4–2. Trends from these data are similar to those reported in children with greater degrees of SNHL. That is, children with minimal degrees of hearing impairment performed poorer than normal hearers across most listening conditions. Moreover, note that the differences in recognition scores between the two groups increased as the listening environment become more adverse. For example, at a S/N ratio of +6 dB, both groups

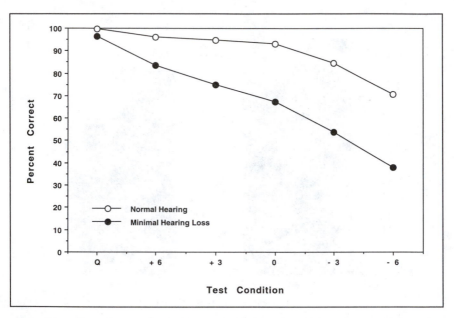

**Figure 4–2.** Mean sentential recognition scores, in percent correct, as a function of signal-to-noise ratio for children with normal hearing sensitivity (indicated by the open circles) and minimally hearing-impaired children (indicated by the closed circles). Figure adapted from Crandell, C. (1993). Speech recognition in noise by children with minimal hearing loss. *Ear & Hearing, 14*(3), 210–216.

obtained recognition scores in excess of 80 percent. At a S/N ratio of –6 dB, however, the minimally hearing-impaired group was able to obtain less than 50 percent correct recognition compared to approximately 75 percent recognition ability for the normal hearers. The speech-recognition difficulties experienced by these children may explain, in part, the psychoeducational and psychosocial deficits often seen in this population (see Bess, 1985).

Several investigators (see Crandell & Smaldino, 1992 for a review) have reported improvements in the psychoeducational development of minimally hearing-impaired children through the use of sound-field amplification. For instance, Sarff (1981) utilized a sound-field amplification system in a classroom with normal-hearing children as well as children with minimal degrees of SNHL. Results indicated that both groups of children, particularly the minimally hearing-impaired children, demonstrated significant improvements in academic achievement when receiving amplified instruction.

# CHILDREN WITH NORMAL-HEARING SENSITIVITY

Perhaps surprisingly, one group of "normal hearers" is younger children with normal-hearing sensitivity. Specifically, investigators have demonstrated that younger listeners require better acoustical environments than do adult listeners to achieve equivalent recognition scores (Crandell, 1992; Crandell, 1993a,b; Elliott, 1979; Elliott, 1982; Elliott, Connors, Kille, Levin, Ball, & Katz, 1979; Nabelek & Nabelek, 1985). Adult-like performance on recognition tasks in noise or reverberation is generally not reached until the child reaches approximately 13–15 years of age.

Based on these data, it is reasonable to assume that commonly-reported levels of classroom noise and reverberation could adversely affect the speech perception of younger children with normal-hearing sensitivity. To support this assumption, let us again examine the Finitzo-Hieber and Tillman (1978) data (see Table 4–2). Note that in typical classroom listening environments, the children with normal hearing generally obtained poor recognition scores. For example, in a relatively good classroom listening environment (S/N ratio = +6 dB; T = 0.4 second), these children were able to recognize only 71 percent of the stimuli. In a poor, but commonly reported classroom environment (S/N ratio = 0 dB; T = 1.2 second), recognition scores were reduced to approximately 30 percent.

Crandell and Bess (1986) examined the speech recognition of young children (5 to 7 years old) with normal hearing in a "typical" classroom environment (S/N ratio =+6 dB; T = 0.45 second). PBK monosyllabic words were recorded through the KEMAR manikin at speaker-listener distances often encountered in the classroom (6, 12, and 24 feet). Results from this investigation are presented in Figure 4–3. As can be noted, there was a systematic decrease in speech-recognition ability as speaker-listener distance increased. Specifically, mean recognition scores of 89, 55, and 36 percent were obtained at 6, 12, and 24 feet, respectively. Overall, these results suggest that normal-hearing children seated in the middle to rear of a typical classroom setting have greater difficulty understanding speech than has traditionally been suspected.

These findings are understandable by examining those of a study by Leavitt and Flexer (1991). In this investigation, the authors utilized the Rapid Speech Transmission Index (RASTI) to estimate speech perception in a classroom. RASTI measurements are based on the hypothesis that noise and reverberation in a room will affect a speech-like signal in ways that can be related to speech perception. Results indicated that in a front row center

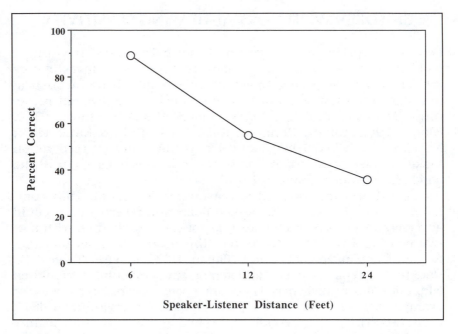

**Figure 4–3.** Mean sentential recognition scores, in percent correct, as a function of speaker-listener distance for children with normal hearing in a "typical" classroom (signal-to-noise ratio = +6 dB; reverberation time = 0.45 second). Figure adapted from Crandell, C., & Bess, F. (1986). Speech recognition of children in a "typical" classroom setting. *ASHA, 29,* 87.

seat of the classroom, only 83 percent of the speech energy was available to the listener. Only 55 percent of the sound energy was available to the listener in the back row center. Clearly, it is reasonable to expect that if only a fraction of the speech signal is available to the listener, poor speech perception will result.

## CHILDREN FOR WHOM ENGLISH IS A SECOND LANGUAGE

Adult listeners for whom English is a second language often experience greater speech-perception difficulties than do native English listeners, particularly in degraded listening environments (Bergman, 1980; Crandell, 1991, 1992; Crandell & Smaldino, 1994; Nabelek & Nabelek, 1985). Bergman (1980), for example, examined the speech perception of adult native-Hebraic listeners under various conditions of acoustic degradations, including noise (signal-to-noise (S/N) ratio= +3 dB), reverberation (RT = 2.5 seconds), and split-band dichotic listening. Results indicated that the non-native English subjects obtained significantly poorer perception scores than did the native-English listeners across all listening conditions. Interestingly, these results were obtained although the native-Hebraic listeners had been English speakers for more than 50 years. Such findings have important educational and therapeutic implications for the more than 2 million non-native English children who may be exposed to unfavorable listening conditions in the classroom and/or resource room environment (Crandell & Bess, 1986, 1987; Crandell, 1991, 1992, 1993a; Crandell & Smaldino, 1992, 1994; Finitzo-Hieber, 1988, Olsen, 1981, 1988; Ross, 1978).

Few investigations have examined the communicative efficiency of non-native English-speaking children under "real world" learning environments. Crandell (1994) examined the speech-perception abilities of 20 native English children and 20 non-native English children under commonly reported classroom S/N ratios (+6 dB, +3 dB, 0 dB, –3 dB, –6 dB). Sentence recognition was assessed by BKB sentences, while the SPIN multibabble was used as the noise competition. Results from this investigation are shown in Figure 4–4. Although both groups obtained equivalent

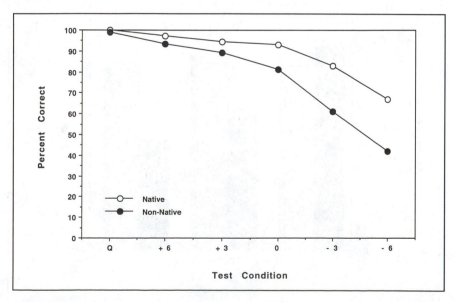

**Figure 4–4.** Mean sentential recognition scores, in percent correct, as a function of signal-to-noise ratio for native English children (indicated by the open circles) and non-native English-speaking children (indicated by the closed circles). Figure adapted from Crandell, C. (1994). The effects of noise on the speech perception of non-native English children. Submitted to *Language, Speech, & Hearing Services in the Schools.*

recognition scores in quiet, the non-native English group performed significantly poorer at S/N ratios ranging from +3 to –6 dB.

Sound field amplification has been shown to benefit the speech perception abilities of non-native English children in the classroom. Crandell (1994) studied the amplified and unamplified speech recognition of native and non-native English children with normal hearing at various speaker-listener distances (6, 12, and 24 feet) in a classroom environment (S/N ratio = +6 dB; T = 0.4 second). Specifically, PBK monosyllabic word lists were recorded through the KEMAR manikin in amplified and unamplified listening conditions for several FM sound-field amplification systems. Results from this investigation (see Figure 4–5) indicated that the speech perception in the non-native English children was significantly improved with the utilization of sound-field amplification at speaker-listener distances of 12 and 24 feet.

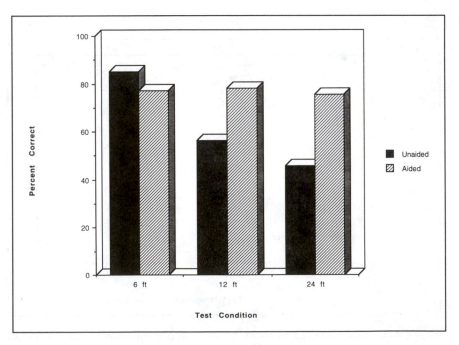

**Figure 4–5.** Speech Perception by non-native English children in a classroom setting (SNR = +6 dB; T = 0.46 second) with and without FM sound-field amplification.

## CHILDREN WITH DEVELOPMENTAL DISABILITIES

Children with developmental disabilities is another population that would benefit, for a number of reasons, from the use of sound-field amplification systems. First, attention to and concentration on tasks, which are basic requirements for learning, usually are deficient in special education populations. Second, there is a greater incidence of conductive hearing impairment among children who experience developmental disabilities than would be expected in a regular classroom population.

Flexer, Millin, and Brown (1990) conducted a study to determine if sound-field amplification could reduce the effects of distractibility, conductive hearing loss, and typical levels of classroom noise in a class for students with developmental disabilities. Several conclusions were drawn from this investigation. First, the nine children who attended a primary-level class for students with developmental disabilities made significantly fewer errors on a word identification task when the teacher presented the words through the sound-field amplification system than when the words were presented without amplification. Second, observation showed the children to be more relaxed and to respond more quickly in the amplified condition. Third, only one of the nine children in the study had normal hearing (15 dB or better at all frequencies in both ears) and normal middle ear function when tested. The incidence of mild hearing impairment among special populations may be underestimated. Even though none of the children in the study was identified as "hearing impaired" according to generally accepted criteria, their minimal to mild hearing losses would interfere with classroom performance.

## CHILDREN WITH CONDUCTIVE HEARING LOSS/RECURRENT OTITIS MEDIA

Approximately 76 to 95 percent of all children experience at least one episode of otitis media with effusion (OME) by six years of age. Additionally, 33 percent of all children develop persistent, or recurrent, OME during their first three years of life; a vital time for speech and language development. Although mild conductive hearing losses of 20 to 30 dB HL are common because of OME, hearing losses as great as 55 dB may be seen in some children (Bess & McConnell, 1981). Even mild losses of hearing can drastically influence the transmission of acoustical information. For example, Dobie and Berlin (1979) examined the potential loss of acoustical information in listeners with simulated losses of only 20 dB. Data indicated that children with mild degrees of conductive hearing loss would experience significant difficulty understanding brief utterances and/or high-frequency stimuli, particularly in degraded listening environments. In addition, these children would lose a majority of transitional information, such as final position consonants and plural endings. Recall from Chapter 3, that this loss of high frequency acoustical information would be significant as research suggests that more than 90 percent of our understanding ability is derived from such phonemes. Discouragingly, research suggests that approximately 10–15 percent of all elementary school age children are experiencing mild hearing losses associated with OME at any given time.

Recurrent OME in children has been linked to compromised speech, language, intellectual, attentional, learning, and psychoeducational and/or psychosocial development. Moreover, a relationship among recurrent OME and reductions in speech-recognition ability has been reported. Gravel and Wallace (1992), for instance, examined the sentence recognition in noise (Pediatric Speech Intelligibility [PSI] Test) of 4-year-old children with positive and negative histories of OME. Speech recognition was assessed with an adaptive procedure with the level of the sentences held constant at 60 dB SPL. Overall, the OME positive children exhibited considerably greater difficulties understanding speech in a noisy environment than did the OME negative group. Specifically, children with a positive history of OME required a significantly greater S/N ratio (2.9 dB) than did the OME negative children to reach equivalent performance levels. Despite the perceptual difficulties of children with conductive hearing loss and children with a history of recurrent OME, there remains a lack of information on the effects of sound-field amplification on these populations.

## CHILDREN WITH ARTICULATION DISORDERS

The most common communication disorder in childhood is errors in speech sound production. Specifically, approximately 5–10 percent of all school-age children exhibit some difficulty in their oral production of speech. Children with articulation difficulties tend to demonstrate poorer speech recognition in noise than do children with normal articulatory abilities (Crandell, Mcquain, &

Bess, 1987; Elliott, 1982). Elliott compared the speech-recognition performance of children with articulation disorders to children with normal articulatory abilities. The SPIN high and low predictability sentence lists were presented binaurally to each child at several S/N ratios (+10, +5, and 0 dB). Although both groups of children obtained essentially 100 percent recognition scores in quiet, the performance of the children with articulation disorders was significantly poorer than in the normal control group. In fact, at a S/N ratio of +5 dB, which is commonly reported in classrooms, children with articulation disorders correctly recognized less than 50 percent of all sentence material presented (across all sentences presented), while the normal control group obtained recognition scores of approximately 65 percent.

## CHILDREN WITH LANGUAGE DISORDERS, CHILDREN WITH CENTRAL AUDITORY PROCESSING DISORDERS, CHILDREN WITH LEARNING DISABILITIES

Speech perception is dependent upon receipt of specific acoustic information, but also upon the ability to relate these often ambiguous cues within the context of a language structure and use of circumstance. The cognitive activities that tie the acoustic signal to language are referred to as learning strategies by McCormick and Schiefelbusch (1984). These cognitive activities, such as selective attention, concept formation, and storage/retrieval, were discussed in Chapter 2. Breakdowns in the processes underlying the cognitive activities will impair the tie between the acoustic signal and language and so produce faulty perception.

Children with learning disabilities, language disorders, and central auditory processing problems all have learning strategies that impair to one degree or another their ability to perceive or use acoustic signals in the classroom. The specific genesis of the problem(s) for each learning strategy dysfunction is well beyond the scope of this chapter. There is, however, one commonality that might be susceptible to improvement through the use of sound-field amplification.

Children who are learning disabled, language disordered, and who have central auditory processing problems frequently display an inability to attend to a desired acoustic signal. Often the inattention is derived from the fact that the desired signal is masked to some degree by surrounding ambient noise, or aspects of the signal simply are not loud enough to reach audibility. Amplifying the acoustic signal to improve the S/N ratio or to provide a completely audible acoustic spectrum might well serve to enable the child to focus on the relevant signal in the classroom and learn to ignore less intense, less relevant acoustic information. Unfortunately, there remains a lack of information regarding the effectiveness of sound-field amplification on these populations of children.

## CHILDREN WITH UNILATERAL SENSORINEURAL HEARING LOSS

Estimates suggest that approximately 2–3 of every 1,000 children exhibit some degree of unilateral SNHL (Bess & Tharpe, 1986; Bess, Tharpe, & Gibler, 1986). Approximately 24–35 percent of these children have failed at least one grade in school. More than 13 percent require special education assistance. The educational difficulties experienced by these children may be explained, in part, by the perceptual difficulties they experience in the classroom. Specifically, children with unilateral SNHL often experience: (1) speech recognition difficulties in noise and reverberation, particularly when speech is presented to the bad ear; and (2) an inability to localize a sound source.

Bess, Tharpe, and Gibler (1986), for example, examined speech perception in 25 children with mild-to-severe degrees of unilateral SNHL. Speech perception was assessed with consonant-vowel (CV) or vowel-consonant (VC) syllables from the Nonsense Syllable test (NST) in several S/N ratios (quiet, +20, +10, 0, and –10 dB). The speech stimuli were presented to the children in two common classroom listening conditions: (1) monaural direct (speech directed at the good ear, noise presented to the bad ear); and (2) monaural indirect (noise presented to the good ear, speech presented to the good ear). While the children with unilateral hearing impairment performed similarly to the normal hearers in quiet, significant differences in perceptual ability were noted between the groups in the noisy listening conditions, particularly in the monaural indirect condition. While it is logical to assume that sound-field systems could significantly improve the speech perception (and academic achievement) of unilaterally hearing-impaired children, there remains an absence of empirical data on this topic.

## ACOUSTICAL STANDARDS FOR CHILDREN WITH "NORMAL HEARING"

Acoustical criteria for appropriate noise levels and reverberation times have not been well established for the diverse populations of "normal hearing" children discussed above. **Until additional research is conducted, a conservative standard for noise levels and reverberation times in listening environments for "normal hearing" children should follow the same acoustical recommendations utilized for hearing-impaired listeners; that is, S/N ratios should surpass +15 dB and reverberation time should not exceed 0.4 second.**

## CONCLUSIONS

This chapter has demonstrated that noise and reverberation levels characteristic of many classroom environments can cause significant reductions in the speech recognition of not only children with hearing impairment, but also of "normal hearers." Certainly, if the accurate recognition of speech is an important educational variable in pediatric listeners, it is imperative that appropriate intervention strategies, such as sound-field amplification, be incorporated into the classroom to assist these populations.

## REFERENCES

Bergman, M. (1980). *Aging and the perception of speech.* Baltimore: University Park Press.

Bess, F. (1985). The minimally hearing-impaired child. *Ear and Hearing, 6,* 43–47.

Bess, F., & McConnell, F. (1981). *Audiology, education and the hearing-impaired child.* St. Louis: C.V. Mosby.

Bess, F.H., & Tharpe, A.M. (1986). An introduction to unilateral sensorineural hearing loss in children. *Ear and Hearing, 7*(1), 3–13.

Bess, F.H., Tharpe, A.M. & Gilber, A. (1986). Auditory performance of children with unilateral sensorineural hearing loss. *Ear and Hearing, 7*(1), 20–26.

Blair, J. (1977). Effects of amplification, speechreading, and classroom environment on reception of speech. *Volta Review, 79,* 443–449.

Blair, J. Myrup, C., & Viehweg, S. (1989). Comparison of the effectiveness of hard-of-hearing children using three types of amplification. *Educational Audiology Monograph, 1*(1),48–55.

Boney, S., & Bess F. (1984). *Noise and reverberation effects in minimal bilateral sensorineural hearing loss.* Paper presented at the American Speech-Language and Hearing Association Convention, San Francisco, CA.

Crandell, C. (1991). Classroom acoustics for normal-hearing children: Implications for rehabilitation. *Educational Audiology Monograph, 2*(1), 18–38.

Crandell, C. (1992). Classroom acoustics for hearing-impaired children. *Journal of the Acoustical Society of America, 92*(4), 2470.

Crandell, C. (1993a). Speech recognition in noise by children with minimal hearing loss. *Ear & Hearing, 14*(3), 210–216.

Crandell, C. (1993b). A comparison of commercially available frequency modulation sound field amplification systems. *Educational Audiology Monograph, 3,* 15–30.

Crandell, C. (1994). The effects of noise on the speech perception of nonnative English children. Submitted to *Language, Speech, and Hearing Services in the Schools*

Crandell, C., & Bess, F. (1986). Speech recognition of children in a "typical" classroom setting. *ASHA, 29,* 87.

Crandell, C., & Bess, F. (1987). Sound-field amplification in the classroom setting. *ASHA, 29,* 87.

Crandell, C., & Schmitt, D. (1994). Sound-field amplification and personal hearing aid use. Paper in preparation to be submitted to *Educational Audiology Monographs.*

Crandell, C., & Smaldino, J. (1992). Sound-field amplification in the classroom. *American Journal of Audiology, 1*(4), 16–18.

Crandell, C., & Smaldino, J. (1994). An update of classroom acoustics for children with hearing impairment. *The Volta Review,* (in press).

Crandell, C., & Smaldino, J. (1994). The importance of room acoustics. In R. Tyler & D. Schum, D. (Eds.), *Assistive listening devices.* Baltimore: William and Wilkins (in press).

Davis, J., Elfenbein, J., Schum, R., & Bentler R. (1986). Effects of mild and moderate hearing impairments on language, educational, and psychosocial behavior of children. *Journal of Speech and Hearing Disorders, 51,* 53–62.

Dobie, R., & Berlin, C. (1979). Influence of otitis media on hearting and development. *Annuals of Otology, Rhinology, and Laryngology, 88,* 48–53.

Elliott, L. (1979). Performance of children aged 9 to 17 years on a test of speech intelligibility in noise using sentence material with controlled word predictability. *Journal of the Acoustical Society of America, 66,* 651–653.

Elliott, L. (1982). Effects of noise on perception of speech by children and certain handicapped individuals. *Sound and Vibration,* December, 9–14.

Elliott, L., Connors, S., Kille, E., Levin, S., Ball, K., & Katz, D. (1979). Children's understanding of monosyllabic nouns in quiet and in noise. *Journal of the Acoustical Society of America, 66,* 12–21.

Finitzo-Hieber, T. (1988). Classroom acoustics. In R. Roeser (Ed.), *Auditory disorders in school children* (2nd ed., pp. 221–233). New York: Thieme-Stratton.

Finitzo-Hieber, T., & Tillman, T. (1978). Room acoustics effects on monosyllabic word discrimination ability for normal and hearing-impaired children. *Journal of Speech and Hearing Research, 21*, 440–458.

Flexer, C., Millin, J., & Brown, L. (1990). Children with developmental disabilities: The effects of sound field amplification in word identification. *Language, Speech, and Hearing Services in the Schools, 21*, 177–182.

Gravel, J., & Wallace, I. (1992). Listening and language at 4 years of age: Effects of early otitis media. *Journal of Speech and Hearing Research, 35*, 220–228.

McCormick, L. and Schiefelbusch, R. (1984). *Early language intervention.* Columbus, OH. Merill Publishing.

Nabelek, A., & Nabelek, I. (1985). Room acoustics and speech perception. In J. Katz (Ed.), *Handbook of Clinical Audiology*, (3rd ed.) Baltimore: Williams & Wilkins.

Olsen, W. (1981). The effects of noise and reverberation on speech intelligibility. In F.H. Bess, B.A. Freeman, & J.S. Sinclair (eds.), *Amplification in education.* Washington, DC: Alexander Graham Bell Association for the Deaf.

Olsen, W. (1988). Classroom acoustics for hearing-impaired children. In F.H. Bess (ed.), *Hearing impairment in children.* Parkton, MD: York Press.

Ross, M. (1978). Classroom acoustics and speech intelligibility. In J. Katz (Ed.), *Handbook of clinical audiology.* Baltimore: Williams and Wilkins

Sanders, D. (1965). Noise conditions in normal school classrooms. *Exceptional Child, 31*, 344–353.

Sarff, L. (1981). An innovative use of free field amplification in regular classrooms. In R. Roeser & M. Downs (Eds.), *Auditory disorders in school children* (pp. 263–272). New York: Thieme-Stratton.

# PART
# B

# PRACTICAL APPLICATIONS OF SOUND-FIELD FM AMPLIFICATION

# CHAPTER
# 5

# ACOUSTIC MEASUREMENTS IN CLASSROOMS

*Joseph Smaldino*
*Carl Crandell*

The classroom has been shown to be a very complex and changing acoustic environment (see Chapter 3). One of the most difficult aspects of measuring or controlling the acoustic environment is that various acoustic parameters change as a function of time. For example, the classroom signal-to-noise (S/N) ratio changes with the intensity of the teacher's voice and the ambient noise in the classroom. The same can be said for (a) the changing speaker-listener distances as the teacher moves around the classroom and (b) to a lesser extent reverberation. **Before changes in the acoustic environment are considered or made, an understanding of the status and dynamics of the particular classroom's acoustics must be known.**

## DETERMINING CLASSROOM ACOUSTIC STATUS

Description of the acoustic environment before application of intervention is important because efficacy of acoustical treatment needs to be documented pre- and post-intervention. Every time classroom acoustic modification is attempted, there is an experiment going on, one in which the outcome of the modification efforts is compared to the pre-modification conditions in order to show that there is an acoustic difference, and that the modification was the primary cause of the change. Demonstration of efficacy of the modification is a primary goal and must be considered a part of the measuring and modification process. Because efficacy is measured by comparing the pre-modification with the post-modification conditions, control over as many of the variables as possible except the modification, will allow cause and effect conclusions to be drawn from the modification. In other words the pre- and post-conditions must be as similar as possible if we are to conclude that the acoustic modification produced the improvement observed. If variables are not controlled, then the cause and effect relationship of the modification will be called into question.

The most straightforward way to control variables is to make the acoustic measurement under real-life conditions (i.e., teacher and students in the room) with normal ambient noises being produced. Because there are many real-life configurations during a school day, the configurations in which most of the instruction takes place must each be considered separately. If this is not done, measurements obtained when students are seated and the teacher is speaking from the front of the room may not be descriptive of when the students are in small groups and the teacher is moving around. The first step in making classroom acoustic measurements, therefore, is to determine the placement of students and teacher when most instruction takes place. As stated earlier, this could be one placement or many. Each must be considered separately.

In order to document the acoustic status of each instructional placement, a small schematic of the real-life conditions is helpful. The distances between the teacher and students should be documented, as well as the distances between individual students. Using a form like that offered in Table 5–1, a detailed schematic should be created for each instructional placement.

## The Acoustic Measurements

Classroom acoustics can be measured with great detail or they can be approximated. Detailed measurements require special instrumentation not readily available in most schools, and so will not be discussed at great length here. Methodologies that produce approximations of classroom acoustics probably provide adequate description of the acoustics, and have the advantage of not requiring very sophisticated instrumentation. The bulk of this section will describe use of approximation methodology.

One of the basic tools used to measure classroom acoustics is the *sound-level meter*. A sound-level meter is a device that measures the amplitude of sound. A listing of companies which manufacture sound-level meters can be found in Appendix A. Sound-level meters range from compact, inexpensive, battery operated

**Table 5–1.** Classroom acoustics documentation form.

---

## Classroom Acoustics Documentation Form

---

Date _____
Teacher _____ Grade _____
Audiologist _____
FM SFA System Used _____

*CLASSROOM SCHEMATIC DIAGRAM*




TEACHER-LISTENER DISTANCE:  Nearest_____feet          Farthest_____feet

TEACHER VOICE LEVEL IN dBA:

| | *Unamplified* | | *Amplified* |
|---|---|---|---|
| Location A | _____ | Location A | _____ |
| Location B | _____ | Location B | _____ |
| Location C | _____ | Location C | _____ |
| Location D | _____ | Location D | _____ |
| Location E | _____ | Location E | _____ |
| Location F | _____ | Location F | _____ |
| Location G | _____ | Location G | _____ |

*REVERBERATION TIME*

Room Volume (V) = _____cubic feet

| | | | |
|---|---|---|---|
| Area Floor | _____X Abs. Coef. _____ | = A Floor | _____ |
| Area Ceiling | _____X Abs. Coef. _____ | = A Ceiling | _____ |
| Area Side Wall 1 | _____X Abs. Coef. _____ | = A Wall 1 | _____ |
| Area Side Wall 2 | _____X Abs. Coef. _____ | = A Wall 2 | _____ |
| Area End Wall 1 | _____X Abs. Coef. _____ | = A End 1 | _____ |
| Area End Wall 2 | _____X Abs. Coef. _____ | = A End 2 | _____ |
| | | Total A | _____ |

RT of classroom = .05 X _____(V) / _____(A) = _____seconds

---

units to computer based devices that can measure numerous properties of an acoustic signal. Sound-level meters are classified according to standards set forth in ANSI S1.14 (1971). Type I meters meet the most rigorous standards, while Type II are general purpose, and Type III are for hobby use. Most serious measurement of classroom noise would require at least a Type II meter. In addition, many sound-level meters incorporate weighting filter networks (see Figure 5–1). The A-weighting network is designed to simulate the sensitivity of the average human ear under conditions of low sound loudness (40 phons). The B-weighting simulates loud sound sensitivity (70 phons), and the C-weighting simulates how the ear would respond to very loud sound. The convention for classroom measurements is the use of the A-weighting network.

Unfortunately, the single number obtained from a sound pressure measurement performed with the A-weighting can be obtained with a number of very different spectra. A more accurate and reliable way to measure spectral intensity would be to do a spectral analysis of the classroom signal and noise. Spectral analysis requires an octave band filter network for the sound-level meter. If an octave band filter network is available, Noise Criteria Curves

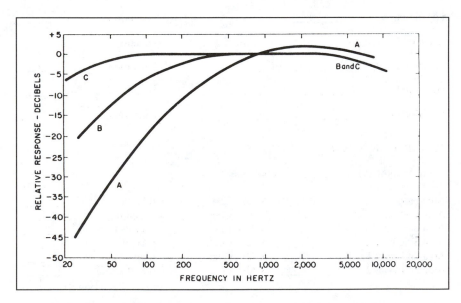

**Figure 5–1.** Weighting scales for sound-level meters. Figure from Berg, F. (1993). *Acoustics and Sound Systems in Schools.* Singular Publishing: San Diego, CA. Used with permission.

(NCC) can be established for a classroom and the suitability of the acoustic spectrum for various human activities can be determined (Beranek, 1954). Noise Criteria Curves are a family of frequency/intensity curves based on octave-band sound pressure across a 20–10,000 Hz band and have been related to successful use of an acoustic space for a variety of activities. The value of each NCC is determined by finding the highest NCC the sound pressure intersects. For instance, Figure 5–2 shows NCCs for normal conversational speech (the highest curve intersected is 55, so the NCC = 55) and a typical occupied classroom environment (NCC = 70). This figure strongly suggests that normal conversational speech would be relatively weak in occupied classroom settings. NCCs have also been roughly equated to sound pressure measures made using the A-weighting. An NCC of 25 is considered suitable for a classroom. The computed equivalent sound pressure level using the A-weighting would be approximately 35 dBA. This would then be the target for the long-term average spectrum in the classroom and may be the best that can be done without extensive spectral analysis as a function of time. It is recom-

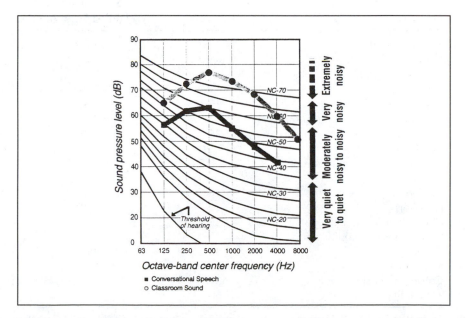

**Figure 5–2.** Noise Criteria Curves for an average occupied classroom setting and conversational speech. Figure from Berg, F. (1993). *Acoustics and Sound Systems in Schools.* Singular Publishing: San Diego, CA. Used with permission.

mended, therefore, that whenever possible, ambient noise levels in classrooms or therapy rooms be measured via NCC measures since this procedure gives the examiner additional information regarding the spectral characteristics of the noise. With this information the audiologist or acoustical engineer can isolate and modify sources of excessive noise in the classroom. Appropriate NCC units, and related A-weighting values, for various communication environments are presented in Table 5–2. Table 5–3 demonstrates the effects of different NCC values on communicative efficiency. An NCC of 20–25 is considered suitable for a classroom. The computed equivalent sound pressure level, using the A-weighting, would be approximately 30–35 dBA. This would be the target sound pressure level for the long-term average spectrum of noise in a classroom.

**Table 5–2.** Appropriate NC units, and related A-weighting values, for various communication environments.

| Type of Space | NC Units | Computed Equivalent SLM Readings (Weighting Scale A, dBA) |
|---|---|---|
| Broadcast studios | 15–20 | 25–30 |
| Concert halls | 15–20 | 25–30 |
| Legitimate theaters (500 seats, no amplification) | 20–25 | 30–35 |
| Music rooms | 25 | 35 |
| Schoolrooms (no amplification) | 25 | 35 |
| Television studios | 25 | 35 |
| Apartments and hotels | 25–30 | 35–40 |
| Assembly halls (amplification) | 25–35 | 35–40 |
| Homes (sleeping areas) | 25–35 | 35–45 |
| Motion-picture theaters | 30 | 40 |
| Hospitals | 30 | 40 |
| Churches (no amplification) | 25 | 35 |
| Courtrooms (no amplification) | 25 | 30–35 |
| Libraries | 30 | 40–45 |
| Restaurants | 45 | 55 |
| Coliseums for sports only (amplification) | 50 | 60 |

Table from Berg, F. (1993). *Acoustics and sound systems in schools.* Singular Publishing: San Diego, CA. Used with permission.

**Table 5–3**. The effects of different NCC values on communicative efficiency.

| NC Units | Communication Environment | Typical Applications |
|---|---|---|
| 20–30 | Very quiet office—telephone use satisfactory—suitable for large conferences | Executive offices and conference rooms for 50 people |
| 30–35 | "Quiet" office; satisfactory for conferences at a 15-ft table; normal voice 10 to 30 ft; telephone use satisfactory | Private or semiprivate offices, reception rooms, and small conference rooms for 20 people |
| 35–40 | Satisfactory for conferencers at a 6- to 8-ft table; telephone use satisfactory; normal voice 6 to 12 ft | Medium-sized offices and industrial business offices |
| 40–50 | Satisfactory for conferences at a 4- to 5-ft table; telephone use occasionally slightly difficult; normal voice 3 to 6 ft; raised voice 6 to 12 ft. | Large engineering and drafting rooms, etc. |
| 50–55 | Unsatisfactory for conferences of more than two or three people; telephone use slightly difficult; normal voice 1 to 2 ft; raised voice 3 to 6 ft | Secretarial areas (typing), accounting areas (business machines), blueprint rooms, etc. |
| More than 55 | "Very noisy"; office environment unsatisfactory; telephone use difficult | Not recommended for any type of office |

Table from Berg, F. (1993). *Acoustics and sound systems in schools.* Singular Publishing: San Diego, CA. Used with permission.

## Classroom Sound Level Measurements

This section will examine how to measure noise in the classroom. The following equipment will be required:

1. sound-level meter (must have A-scale and slow response);
2. 20 ft measuring tape;
3. standard reading passage.

**Step 1.** Position the teacher in the normal instructional position in the classroom. The students should be seated in their normal seats for instruction. It is important that the measurements are made in the time period when instruction normally occurs, so that the acoustic conditions are representative of actual instructional environments.

**Step 2.** Turn on the sound level meter; be sure it is set on the A-weighting scale and on slow response. If you can set the range of the meter, set it to accommodate 40–60 dB SPL to begin.

**Step 3.** Position the sound-level meter to approximate the center of each selected student's head while he/she is seated at his/her desks. Point the sound-level meter toward the teacher position, taking care to avoid placing your body in the sound path between teacher and student, which can produce inaccurate measurements. Student desks at the four corners, the middle, and the middle back of the classroom seating should be measured. More locations can be measured if desired.

**Step 4.** With the students quiet, measure the ambient noise level at the selected student locations and record it on the classroom documentation form. If the noise level fluctuates, take three measurements at 1-minute intervals and average the readings and record those on the form. These measurements will provide an estimate of the ambient noise level during an instructional period. If measurements can only be taken when the students are not in the classroom (i.e., when it is unoccupied), you may convert the unoccupied noise levels to occupied by adding 10 dB to each unoccupied measurement. This conversion is roughly equal to the known difference in noise level between *average* unoccupied and occupied classrooms.

**Step 5.** The teacher should begin reading the standard reading passage at a normal instructional intensity level.

**Step 6.** Repeat Step 4, now that the teacher is reading the standard passage, and record the teacher signal levels on the form. These measurements provide an estimate of the signal level during an instructional period.

**Step 7.** Subtract the ambient noise measurement from the teacher voice measurement to determine the S/N ratio of the classroom at the selected student sites. For example, a student location with a teacher voice level of 60 dBA and a noise level of 50 dBA would have a S/N ratio of +10 dB. One with a teacher level of 60 dBA and a noise level of 70 dBA would have a S/N of −10 dB.

## Classroom Reverberation Measurements

This section will examine how to measure reverberation in the classroom. The following equipment will be required:

1. 20 ft measuring tape or ultrasonic distance estimator;
2. calculator.

The formula used to estimate classroom reverberation time (RT) is:

$$RT = .05 \ V/A,$$

where:    $RT$ = reverberation time in seconds
          $.05$ = a constant
          $V$ = the volume of the room
          $A$ = total absorption of the room surfaces in Sabins

**Step 1.** All of the reverberation estimates can be conducted in an unoccupied classroom. Because a formula is used, no improvement in accuracy is obtained with students and teacher present. During more detailed measurements, the presence of the room occupants would be desirable.

**Step 2.** Calculate the volume of the classroom. This is done by measuring the length, the width, and the height of the classroom in feet and multiplying them together (volume = length of room × width of room × height of room). Record the resultant room volume in cubic feet on the classroom documentation form.

**Step 3.** Multiply the volume of the room by the constant .05 to obtain the numerator for the **RT = .05 V/A** equation. Record the results on the classroom form.

**Step 4.** In order to obtain the denominator of the equation, the area of the walls, floor, and ceiling of the room must first be calculated in square feet. If the walls, ceiling, or floor are irregularly shaped, each section must be measured separately. The area of the floor and ceiling is determined by multiplying the length of the floor or ceiling times its width. The area of the walls can be obtained by multiplying the length of each wall by its height. Enter the values for the area of each (floor, ceiling, walls) on the classroom documentation form.

**Step 5.** The absorption coefficient (Abs. Coef.) is a measure of the sound reflectiveness of different construction materials. The coefficient, expressed in Sabins, must be determined for the material composing the walls, ceiling and floor. Average absorption coefficients are given in Table 5–4 for the most common construction materials. If a different construction material is encountered and you use another absorption coefficient table, average the coefficients given in the other table for 500, 1000, and 2000 Hz for the purposes of these calculations. Enter the average absorption coefficient in the appropriate place on the documentation form.

**Step 6.** Multiply the area of each floor, ceiling, and wall times the absorptive coefficient of the material composing the surface. Add up all of the resultants of the multiplications to obtain the A (total absorption of the room in Sabins) in the **RT = .05 V/A** formula for the room, and record it on the form.

**Step 7.** Take the numerator from Step 3 (.05 × V) and the denominator from Step 6 (A = total absorption in Sabins for the

**Table 5–4**. Sound absorption coefficients for common classroom materials.

| Material— Walls | Average Absorption Coefficient | Material— Floors | Average Absorption Coefficient | Material— Ceilings | Average Absorption Coefficient |
|---|---|---|---|---|---|
| brick | 0.04 | wood parquet on concrete | 0.06 | plaster, gypsum, or lime on lath | 0.05 |
| painted concrete | 0.07 | linoleum | 0.03 | acoustic tiles (5/8"), suspended | 0.68 |
| window glass | 0.12 | carpet on concrete | 0.37 | acoustic tiles (1/2"), suspended | 0.66 |
| plaster on concrete | 0.06 | carpet on foam padding | 0.63 | acoustic tiles (1/2"), not suspended | 0.67 |
| plywood | 0.12 | | | high absorptive panels suspended | 0.91 |
| concrete block | 0.33 | | | | |

Table adapted from Berg, F. (1993). *Acoustics and sound systems in schools*. Singular Publishing: San Diego, CA. Used with permission.

room) and divide them in order to determine the estimated reverberation time of the room in seconds (**RT = .05 V/A**). Enter the estimate on the documentation form. It should also be noted that the RT of a room can be obtained using a reverberation meter. A listing of companies which manufacture RT meters can be found in Appendix A.

The classroom documentation form should now have recorded on it, the geometry of the classroom for instruction, the S/N ratio of the classroom at selected student sites during normal instruction, and the reverberation time of the room. These measurements form the current acoustic status of the room and are the basis from which recommendations for classroom improvements will be made.

Prior to discussing classroom recommendations, it should be noted that there are several statistical indices which can also assist in the determination of acoustics in a classroom. One such method, the Rapid Speech Transmission Index (RASTI), is often used to estimate speech perception in a classroom. RASTI is based on the assumption that noise and reverberation in a room will affect a speech-like signal in ways that can be related to speech perception. A modulation transfer function is derived as a speech-like signal is transmitted and revived in a room. The reader is directed to Berg (1993) for information on additional procedures, such as RASTI and ALcons, to measure acoustical environments.

# CLASSROOM RECOMMENDATIONS

**Step 1.** Compare classroom acoustic results to the recommended classroom standards summarized in Table 5–5. If the classroom meets the standards or exceeds them, no acoustic modification is necessary. If the standards are not met, however, acoustic modification should be attempted and/or;

**Step 2.** Since speaker-listener distance affects both S/N and RT, an inspection of the classroom geometry on the classroom documentation form may reveal that a reduction of the distance is possible and should be tried. Since a halving of the speaker-listener distance can be expected to increase the signal level by 6 dB as predicted by the Inverse Square Law and/or;

**Step 3.** Adverse RT can be affected by structural changes in the room such as adding carpet, placing acoustic ceiling tile in the room, and the like. Structural modifications are described elsewhere (see Chapter 6) and will not be repeated here. Typically, structural changes are cost prohibitive and are not frequently used to improve classroom acoustics and/or;

**Step 4.** Install a sound-field FM amplification system in the classroom. The details of selection and installation are covered in detail in Chapter 9. The goal of sound-field installation is to increase the signal strength by about 10 dB. The rationale for 10 dB was presented in Chapter 1, and/or;

**Step 5.** Given that you have intervened in ways described in Steps 1–4, the task is to determine if the intervention accomplished the anticipated goal. Remember, the goal is simply to increase the level of the teacher's voice 10 dB uniformly throughout the classroom measurement locations, and/or;

**Step 6.** Included on the classroom documentation form under "Teacher voice level" is a section designated "Amplified." Simply repeat the teacher voice level measurements with the intervention in place and record the new voice levels. If they are 10 dB or greater at each location, the goal has been accomplished.

**Table 5–5.** Recommended classroom acoustic standards.

| | |
|---|---|
| Ambient Noise Level | 30–35 dBA |
| Noise Criteria Curve | 20–25 |
| Reverberation Time | 0.4 second |
| Signal-Noise Ratio | +15 dB |

If the levels do not attain the goal, modification or additional intervention is necessary. Upon completion of the addition and/or modification, repeat the measurements to ascertain goal attainment. This cycle is repeated until the goal has been reached.

# REFERENCES

American National Standards Institute. (1971). Specification for sound level meters. ANSI S1.14-1971: New York.

Beranek, L. (1954). *Acoustics*. New York: McGraw-Hill.

Berg, F. (1993). *Acoustics and sound systems in schools*. San Diego, CA: Singular Publishing.

# APPENDIX A
# SOURCES FOR SOUND LEVEL-REVERBERATION METERS

**Bruel & Kjaer Instruments**
185 Forest
Marlborough, MA 01752-3093
(508) 481-7737

**Communications Company**
3490 Noell Street
San Diego, CA 92110
(619) 297-3261

**Goldline**
P. O. Box 500
West Reading, CN 06896
(203) 938-2588

**Larson-Davis Laboratories**
1681 West 820 North
Provo, UT 84601
(801) 375-0177

**Quest Electronics**
510 Worthington
Oconomosoc, WI 53006
(800) 245-0779

**Realistic Corporation**
A division of Radio Shack-
    Tandy Corporation
P. O. Box 1052
Fort Worth, TX 76102
(800) 390-3011

**Techron**
1718 W. Mishawaka Rd.
Elkhart, IA 46517
(219) 294-8300

# CHAPTER
# 6

# ACOUSTICAL MODIFICATIONS IN CLASSROOMS

*Carl Crandell*
*Joseph Smaldino*

The speech-recognition difficulties experienced by children with hearing impairment and children with "normal hearing" highlight the need to provide an appropriate listening environment for these populations. **Recall from Chapter 3 that acoustical standards for such populations indicate that signal-to-noise (S/N) ratios should exceed +15 dB; unoccupied noise levels should not exceed 30–35 dB(A), while reverberation times should not surpass 0.4 second.** As noted, however, these acoustical recommendations are rarely achieved in "real-world" listening environments. This chapter will examine various modifications of the acoustical environment to facilitate the speech perception of pediatric listeners in educational environments. The reader is directed to the following sources (Beranek, 1954; Bess & McConnell, 1981; Crandell, 1991, 1992; Crandell & Smaldino, 1994a, b; Finitzo-Hieber, 1988; Finitzo-Hieber & Tillman, 1978; Knudsen & Harris, 1978; Niemoller, 1968; Olsen, 1977, 1981, 1988; Ross, 1978) for additional discussions pertaining to acoustical modifications in the classroom.

## REDUCTION OF NOISE AND REVERBERATION LEVELS IN THE CLASSROOM

Recall from Chapter 3 that ambient classroom noise can originate from several possible sources. These sources include *external noise* (noise that is generated from outside the school); *internal noise*

(noise that originates from within the school building, but outside the classroom); and *classroom noise* (noise that is generated within the classroom). In order to conduct the most appropriate modification of the classroom, it must first be determined which specific noise source, or sources, needs to be reduced. Moreover, the reverberant characteristics of the enclosure must be quantified. With these considerations in mind, the remainder of this chapter will outline various procedures in order to reduce noise and reverberation levels in the classroom. Acoustical modification of the classroom must be conducted prior to sound-field utilization.

## Reduction of External Noise Levels

**1.** Classrooms utilized for children who are hearing impaired, or for "normal hearers", must be located away from high noise sources, such as busy automobile traffic, railroads, construction sites, airports, and furnace/air conditioning units. The most effective procedure for achieving this goal is through appropriate planning with contractors, school officials, architects, architectural engineers, audiologists, and teachers for the hearing impaired **before** the design and construction of the school building. Such consultation should include strategies for locating classrooms from high external noise sources. Moreover, acoustical modifications such as the placement of vibration reduction pads underneath the supporting beams of the building to reduce structure-borne sounds can be implemented. Sadly, consultation among such disciplines prior to building construction is rare (Crandell & Smaldino, 1994a).

**2.** *A Sound Transmission Loss (STL)* of at least 45 to 50 dB is required for external walls. Sound Transmission Loss refers to the amount of noise that is attenuated as it passes through a particular surface. For instance, if an external noise of 100 dB SPL was reduced to 60 dB SPL in the classroom, the exterior wall of that room would have a STL of 40 dB SPL. Sound Transmission Losses of common building materials are presented in Table 6–1. Note that a seven-inch concrete wall provides approximately 53 dB attenuation of outside noise, while windows and doors provide only 24 dB and 20 dB attenuation, respectively. Therefore, doors and/or windows on the external wall should be avoided in situations of high external levels. Additional procedures to increase the STL of an external wall include: (1) the placement of absorptive materials (such as fiberglass material) between the wall studs; (2) thick or double concrete construction on the exterior wall; and (3) the addition of several layers of gypsum board (at least 5/8") or plywood material.

Table 6–1. Sound transmission losses, in dB, for various building materials.

| Materials | Total Thickness | | Frequency (Hz) | | | | | | STL rating |
|---|---|---|---|---|---|---|---|---|---|
| | inch | cm | 125 | 250 | 500 | 1k | 2k | 4k | |
| **Walls** | | | | | | | | | |
| Solid concrete | 3 | 8 | 35 | 40 | 44 | 52 | 59 | 64 | 47 |
| Concrete (6;15), layers of plaster | 7 | 18 | 39 | 42 | 50 | 58 | 64 | 66 | 53 |
| Solid concrete blocks, layers of plaster | 16 | 41 | 50 | 54 | 59 | 65 | 71 | 68 | 63 |
| Brick (4½;11), layers of plaster | 5½ | 14 | 34 | 34 | 41 | 50 | 56 | 58 | 42 |
| Brick (9'23), layers of plaster | 10 | 25 | 41 | 43 | 49 | 55 | 57 | 59 | 52 |
| Stone (24;61), layers of plaster | 25 | 64 | 50 | 53 | 52 | 58 | 61 | 68 | 56 |
| Hollow concrete block | 6 | 15 | 32 | 33 | 40 | 48 | 51 | 48 | 43 |
| Cinder block (4;10), layers of plaster | 5¼ | 13 | 36 | 37 | 44 | 51 | 55 | 62 | 46 |
| Hollow gypsum block (3;8), layers of plaster | 4 | 10 | 39 | 34 | 38 | 43 | 48 | 46 | 40 |
| Double brick (4½;11) wall, cavity (2;5), layer of plaster | 12 | 31 | 37 | 41 | 48 | 60 | 60 | 61 | 49 |
| Double brick (4½;11) wall, cavity (6;15), layer of plaster | 18 | 46 | 48 | 54 | 58 | 64 | 69 | 75 | 62 |
| Solid sanded gypsum plaster | 2 | 5 | 36 | 28 | 35 | 39 | 48 | 52 | 36 |
| Solid gypsum core moveable partition | 2¼ | 6 | 34 | 34 | 37 | 38 | 39 | 45 | 36 |
| **Floor-ceiling** | | | | | | | | | |
| Reinforced concrete slab | 4 | 10 | 48 | 42 | 45 | 55 | 57 | 66 | 44 |
| Reinforced concrete as above + carpeting and pad | 4½ | 11 | 48 | 42 | 45 | 55 | 57 | 66 | 44 |
| Concrete (4½;11), wood flooring, layer of plaster | 7 | 18 | 35 | 37 | 42 | 49 | 58 | 62 | 46 |
| Concrete (4½;11), screed, suspended plaster ceiling | 10 | 25 | 38 | 41 | 45 | 52 | 57 | 59 | 48 |

*(continued)*

**Table 6–1.** *(continued)*

| Materials | Total Thickness inch | cm | Frequency (Hz) 125 | 250 | 500 | 1k | 2k | 4k | STL rating |
|---|---|---|---|---|---|---|---|---|---|
| Floor-ceiling *(continued)* | | | | | | | | | |
| Concrete (6;15), wood, battens floating on glass wool, layer of plaster | 9 1/2 | 24 | 38 | 44 | 52 | 55 | 60 | 65 | 55 |
| Wooden joists (8;20), floor gypsum wallboard | 9 1/2 | 24 | 19 | 24 | 31 | 35 | 45 | 42 | 34 |
| Wooden joists (7;18), wood+linoleum, reeds+plaster | 9 1/2 | 24 | 24 | 27 | 35 | 44 | 52 | 58 | 39 |
| Windows | | | | | | | | | |
| Double window | 3/8 | 1.0 | 21 | 22 | 19 | 24 | 25 | 33 | 24 |
| Double window sealed | 3/8 | 1.0 | 20 | 25 | 20 | 30 | 34 | 34 | 28 |
| Double window with cracks | 3/8 | 1.0 | 18 | 21 | 19 | 20 | 22 | 30 | 20 |
| Doors | | | | | | | | | |
| Solid core wood, weather strip | 1 3/4 | 4.0 | 21 | 27 | 30 | 26 | 25 | 29 | 27 |
| Hollow core wood, weather strip | 1 3/4 | 4.0 | 14 | 15 | 17 | 18 | 22 | 29 | 20 |

Adapted from Lipscomb, D. (1978). *Noise in audiology*. Austin, TX: Pro-Ed.

**3.** All exterior walls must be free of cracks or openings that would allow extraneous noises into the classroom. Even small openings in external walls can significantly reduce the STL.

**4.** If windows are located on the external wall, they must be properly installed, heavy weighted or double-paned (such as storm windows), and should remain closed. During numerous acoustical analyses of classrooms, the authors have unfortunately seen exterior windows that remain open the majority of the school day. In addition, existing windows can be sealed with non-hardening caulk to increase the STL. Of course, safety regulations must be checked before sealing outside windows.

**5.** Landscaping strategies can also attenuate external noise sources. These strategies include the placement of: (a) trees or shrubs (that bloom all year long), and (b) earthen banks around the school building.

**6.** Finally, solid concrete barriers with an STL of 30–35 dB can be placed between the school building and the noise source to reduce external noise entering into the classroom.

## Reduction of Internal Noise Levels

**1.** The most cost-effective procedure for reducing internal noise levels in the room is to relocate the classroom to a quieter area of the building. Classrooms must not be located next to a high noise source such as the gymnasium, metal shop, cafeteria, or bandroom. At least one quiet environment, such as a storage area or closet, should separate classrooms from each other or from high noise sources in the school building. If the classroom cannot be relocated, then instruction areas need to be relocated away from the internal noise source, such as an adjacent classroom.

**2.** If suspended ceilings separate the classroom from another room, then fiberglass installation or lead sheets should be placed in the plenum space above the wall.

**3.** Double or thick wall construction should be used for the interior walls, particularly those walls that face noisy hallways or rooms. Recall that additional layers of gypsum board, plywood, and/or the placement of absorptive materials between wall studdings can also increase the attenuation characteristics of interior wall surfaces. Moreover, all cracks between classrooms need to be fixed.

**4.** Acoustical ceiling tile and/or carpeting should be used in hallways outside the classroom (Please see Appendix A for a listing of several companies that manufacture acoustically treated materials.)

**5.** All classrooms should contain acoustically treated or well-fitting (preferably with rubber or gasket seals) high mass per unit area doors. Hollow-core doors between rooms, or facing the hallway, should not be utilized. Doors (or interior walls) should not contain ventilation ducts that lead into the hallways.

**6.** Heating or cooling ducts that serve more than one classroom need to be lined with acoustical materials or furnished with baffles to decrease noise emitting from one room to another.

**7.** Permanently mounted blackboards can be backed with absorptive materials to reduce sound transmission from adjacent rooms.

## Reduction of Classroom Noise Levels

**1.** The simplest procedure to reduce the effects of classroom noise is to position children away from high noise sources, such as fans, air conditioners, or heating ducts, faulty lighting fixtures, and doors or windows that lead to high noise sources. Often, however, classroom noise sources are so intense that no location in the classroom is appropriate for adequate communication. In these cases, acoustical modification must be conducted.

**2.** Malfunctioning air conduction/heating units and ducts need to be replaced or acoustically treated. Heating ducts, for example, can be lined with acoustical materials to reduce both vibratory and air-borne noise. In addition, rubber supports and flexible sleeves or joints should be used to reduce the transmission of structural-borne noise through the ductwork system. Moreover, all fans and electrical motors in air conditioning/heating units must be lubricated and maintained on a regular basis.

**3.** Installation of thick, wall-to-wall carpeting (with adequate padding) to dampen the noise of shuffling of hard-soled shoes, the movement of desks/chairs, and so on can also reduce classroom noise levels.

**4.** Acoustical paneling can be placed on the walls and ceiling. Wall paneling typically should be placed partly down the wall and not on walls parallel to one another.

**5.** The placement of some form of rubber tip (tennis balls cut in half work well) on the legs of desks and chairs can decrease classroom noise. This recommendation is particularly important if the classroom is not carpeted.

**6.** Acoustically treated furniture can be purchased for classrooms. although it must be noted that such furniture can be expensive and may present hygiene problems.

**7.** Hanging of thick curtains or acoustically treated venetian blinds over window areas to dampen classroom noise levels can be effective.

**8.** Avoid open-plan classrooms particularly for children with hearing impairment as it is well recognized that such classrooms are considerably noisier than regular classrooms.

**9.** Instruction should not take place in areas separated from other teaching areas by sliding doors, thin partitions, and/or temporary walls. Walls between instruction areas must be of sufficient thickness and contiguous between the solid ceiling and floor. Walls that are contiguous only to a suspended false ceiling allow for significant sound transmission between rooms.

**10.** Fluorescent lighting systems, including the ballast, need to be regularly maintained and replaced if faulty.

**11.** Typewriter or computer keyboard noise can be lowered by the placement of rubber pads, or carpet remnants, under such instruments. Whenever possible, such instruments (as well as any other office equipment) should be located in separate rooms. Rubber pads to reduce vibratory noise should be placed under all office equipment in the school.

**12.** Children can be encouraged to wear soft-soled shoes.

## Reduction of Classroom Reverberation

**1.** Reverberation can be reduced by covering the hard reflective surfaces in a classroom with absorptive materials, such as acoustical paneling. Certainly, to reduce reverberation, ceilings should be covered with acoustical paneling. Acoustical panels may also be placed on walls, but typically not on walls parallel to one another. Cork bulletin boards, carpeting, and bookcases can also be strategically placed on the walls, however such materials are not as absorptive as acoustical paneling. Interestingly, the installation of absorptive materials will not only reduce reverberation in the environment, but also will decrease the noise level in the room by 5–8 dB.

**2.** Thick carpeting on the floors can also significantly reduce reverberation. Classrooms that contain both ceiling tile and carpeting have approximately 60 percent of room surfaces covered with absorptive material.

**3.** Curtains or thick draperies need to be placed to cover the hard reflective surfaces of windows. Even when the curtains are open, they will serve to reduce the reverberation time of the enclosure.

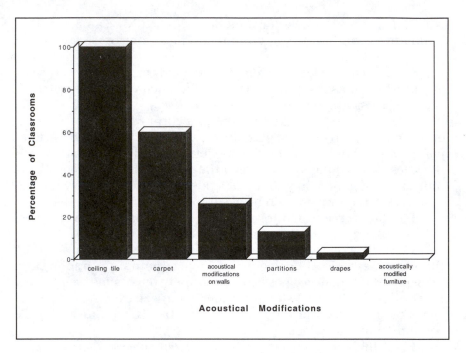

**Figure 6–1.** Acoustical modifications in 32 classrooms for the hearing impaired. Figure used with permission from Crandell, C., & Smaldino, J. (1994a). An update of classroom acoustics for children with hearing impairment. *The Volta Review*, (in press).

**4.** Positioning of mobile bulletin boards and blackboards at angles other than parallel to opposite walls will also reduce the reflected sound in an enclosure.

**5.** Some teachers have used creative artwork from egg cartons or carpet scraps attached to walls or suspended from ceilings to help absorb noise and reduce reverberation.

## Current Status of Acoustical Modifications in Classrooms

**Despite the numerous procedures for treating the acoustical environment, classrooms often exhibit minimal degrees of acoustical modifications.** Bess, Sinclair, and Riggs (1984) reported that while 100 percent of classrooms had acoustical ceiling tile, only 68 percent had carpeting, and only 13 percent had draperies. None of the classrooms contained any form of acoustical furniture treatment. Crandell and Smaldino (1994a) reported that while all of 32 classrooms examined had acoustic ceiling tile, only 14 (54 percent) contained carpeting. Moreover, only one of the classrooms had drapes, while none of the rooms had acoustical furniture treatments (see Figure 6–1).

# REFERENCES

Beranek, L. (1954). *Acoustics.* New York: McGraw-Hill.

Bess, F., & McConnell, F. (1981). *Audiology, education and the hearing-impaired child.* St. Louis: C.V. Mosby.

Bess, F., Sinclair, J., & Riggs, D. (1984). Group amplification in schools for the hearing-impaired. *Ear and Hearing, 5,* 138–144.

Crandell, C. (1992). Classroom acoustics for hearing-impaired children. *Journal of the Acoustical Society of America, 92*(4), 2470.

Crandell, C., & Smaldino, J. (1994a). An update of classroom acoustics for children with hearing impairment. *The Volta Review,* (in press).

Crandell, C., & Smaldino, J. (1994b). The importance of room acoustics. In R. Tyler, & D. Schum (Eds.), *Assistive listening devices.* Baltimore: Williams and Wilkins (in press).

Finitzo-Hieber, T. (1988). Classroom acoustics. In R. Roeser, & M. Downs (Eds.), *Auditory disorders in school children* (2nd ed., pp. 221–233). New York: Thieme-Stratton.

Finitzo-Hieber, T., & Tillman, T. (1978). Room acoustics effects on monosyllabic word discrimination ability for normal and hearing-impaired children. *Journal of Speech and Hearing Research, 21,* 440–458.

Knudsen, V., & Harris, C. (1978). *Acoustical designing in architecture.* Washington, DC: The American Institute of Physics for the Acoustical Society of America.

Niemoller, A. (1968). Acoustical design of classrooms for the deaf. *American Annals of the Deaf, 113,* 1040–1045.

Olsen, W. (1977). Acoustics and amplification in classrooms for the hearing impaired. In F.H. Bess (Ed.), *Childhood deafness: Causation, assessment and management.* New York: Grune & Stratton.

Olsen, W. (1981). The effects of noise and reverberation on speech intelligibility. In F.H. Bess, B.A. Freeman, & J.S. Sinclair (Eds.), *Amplification in education.* Washington, DC: Alexander Graham Bell Association for the Deaf.

Olsen, W. (1988). Classroom acoustics for hearing-impaired children. In F.H. Bess, (Ed.), *Hearing impairment in children.* Parkton, MD: New York.

Ross, M. (1978). Classroom acoustics and speech intelligibility. In J. Katz (Ed.), *Handbook of clinical audiology.* Baltimore: Williams and Wilkins.

# APPENDIX A
# SOURCES FOR ACOUSTICAL MATERIALS

**Armstrong**
Suite 115
O'Hare Lake Office Plaza
2300 E. Devon Avenue
Des Plaines, IL 60018
(312) 299-9030

**Illbruck/USA**
3800 Washington Avenue
Minneapolis, MN 55412
(612) 521-3555

**Industrial Acoustics**
1160 Commerce Avenue
Bronx, NY 10462
(212) 931-8000

**Kinetics Noise Control**
6300 Irelan Place
P. O. Box 655
Dublin, OH 43017
(614) 889-0480

**Kinetics West**
7059 South Curtice
Littleton, CO 80120
(303) 797-0273

**Metals Building Interior**
Products Company
5309 Hamilton Avenue
Cleveland, OH 44144
(216) 431-6400

**Proudfoot**
P. O. Box 338
Botsford, CT 06404-0338
(203) 459-0033
(800) 445-0034

**RPG Diffusor System**
373 W. 651
Commerce Drive
Upper Marlboro, MD 20772
(301) 249-0044

**Sound Concepts**
2810 Urbandale Lane
Plymouth, MN 55447
(612) 473-1454

**Sound Reduction Corporation**
16601 St. Clair Avenue
Cleveland, OH 44110
(216) 481-1900

**Systems Development Group**
5744 Industry Lane
Suite J
Fredrick, MD 21701
(800) 221-8975
(301) 846-7990

**Technicon**
4412 Republic Drive
Concord, NC 28027
(704) 788-1131

**United Process**
279 Silver
P. O. Box 545
Agawam, MA 01001
(413) 789-1770

# CHAPTER
## 7

# IDENTIFYING AND MANAGING THE LEARNING ENVIRONMENT

*Carolyn Edwards*

Thehe use of personal FM systems generally affects only two individuals—the sender (typically the teacher) and the receiver (the child with hearing loss). Sound-field FM systems, however, affect the entire class. The teacher continues to use the transmitter but now one or more loudspeakers send a diffuse signal throughout the entire classroom. **Audiologists must therefore consider the entire learning environment in the installation, maintenance, and acceptance of sound-field FM systems by teachers and students**. This chapter highlights some of the variables of children's learning environments relevant to the use of sound-field FM systems. These variables include classroom style, classroom design, learning style, and teacher receptivity.

## CLASSROOM STYLE

A variety of classroom styles are used in the educational setting, including closed classrooms, open concept, split grades, and team teaching.

### Closed Classroom

The most traditional format is the **closed classroom** in which the teacher and students are in a completely enclosed space; that is, separated from other classes. This situation is the least complex installation for sound-field FM systems, since only one teacher's activity schedule must be considered.

## Open Concept Areas

**Open concept areas** were more popular in the eighties, but continue to be used by a number of school boards. The purpose of open concept classrooms is to permit team teaching, cross groupings of children between classes, and movement of children between classrooms without physical barriers. Thus, as many as eight or more classes could be located in the same space without any walls to separate the classrooms. Enhancement of the speech signal is highly desirable since the speech-to-noise (S/N) ratios are often poorer than in closed classrooms for two reasons: (1) the background noise levels are predictably higher, and (2) the teachers often use a softer than average voice level in order to reduce interference with teachers in adjacent classes.

The use of sound-field FM systems in open concept environments is quite challenging, and may be impossible in some sites. Teachers in open concept areas who use sound-field FM systems need to coordinate schedules with teachers in adjacent classes so that they are not using the sound-field system at the same time. Otherwise, the enhanced signal from one class can interrupt students listening to another teacher on another frequency in an adjacent area. Classes at opposite ends of the open teaching areas, however, may be able to utilize amplification without any problems. Given the necessity to modify schedules and work together, the teachers' willingness to experiment is essential for successful introduction of sound-field FM systems in open-plan settings.

## Split-Grade Classrooms

When there is an insufficient number of students to form a class of a single grade, schools often combine two grades in one classroom with a single teacher. Teachers in **split-grade classrooms** tend to teach one grade while the other grade is working on small group or independent activities. It is difficult to amplify the teacher's voice during a lesson to Grade Four students and not distract Grade Five students who are working on their own. It is therefore desirable to use sound-field FM systems where the teacher can carry the speaker(s) to the area where she/he will be teaching in the room, or use systems where speakers can be turned off in the part of the room in which she/he is not instructing.

## Team Teaching

Two teachers will sometimes combine classes together in one room and **team teach**. Given the availability of only one microphone on a specified frequency in each room, the teachers using the sound-field FM system in this situation are forced to pass the transmitter back and forth, or choose one teacher who will transmit through the system. Some companies can modify the FM transmitter to accept a pass-around microphone; then one teacher can wear the transmitter and the other teacher can talk through the pass-around microphone patched into the transmitter via a 12–15 foot cord. Although this is a wired option that necessitates being in close proximity with each other, use of the pass-around microphone is the only means of amplifying both teacher's voices with ease.

## CLASSROOM DESIGN

### Portable Classrooms

The type of classroom and the use of classroom space affect the use and placement of sound-field FM systems. In schools where the enrollment has far exceeded the original capacity of the school, **portable classrooms** have been installed. Portables are stand-alone installations outside the school building, with separate ventilation and heating systems. Unfortunately, the hollow floors designed to enhance portability create greater reverberation than a solid floor, and often the rooms are uncarpeted, leading to increased noise levels from the sounds of chairs and desks scraping on the floor. The author measured an intensity level of 84 dB(A) from the sound of a single chair scraping on the floor of a portable classroom. Naturally, teachers tend to use louder voices to compensate for the poorer acoustic conditions, resulting in more rapid vocal fatigue than that experienced by teachers within the school building. Teachers, particularly in portable classrooms, notice an immediate decrease in vocal intensity with the introduction of a sound-field FM system, and a resulting decrease in vocal fatigue.

## Classroom Space Utilization

The **use of classroom space** reflects the teaching style of the individual teacher. Some will spread the students' desks throughout the room, others will group the desks together and use the remaining space for activity centers and storage, and others will set aside some space for children to work, read, or play on their own. It is more difficult to install speakers in a classroom partway through the school year when the teacher has already designed the layout of the class. Meeting with the classroom teacher at the beginning of the year allows the audiologist to become familiar with the classroom traffic flow. In addition, it gives the teacher the opportunity to make changes in floor plans that increase the ease of speaker installation. Speakers should not be located in "dead space" where no instruction occurs, or close to areas set aside for quiet time where children go to avoid the bustle of classroom activities.

## Classroom Seating

The **type of classroom seating** adopted by the teacher is determined by teaching style and age of the students. In the younger grades, where interactive learning approaches are favored, teachers create up to six activity centers in the room where children rotate in and out in small groups. Children generally circle around the teacher for large group activities. The ideal amplification for activity centers is a portable speaker that the teacher can carry. An option for multiple speaker systems is an off/on toggle switch installed in the wiring just behind the speaker. In this setup, the teacher can turn the speaker on when talking to children in one activity center, and turn the speaker off when leaving the activity center so that children in other activity centers do not receive the amplification. Large group instruction with young children often occurs within only one part of the classroom; positioning of single or multiple speaker systems must permit satisfactory amplification of large group instruction in addition to the activity centers. In the older grades where didactic approaches are more common, row seating is typical and there are less variables to consider in speaker placement (see Chapter 9).

## LEARNING STYLE

The need for a sound-field FM system is dictated, in part, by the learning styles favored by the classroom teacher. Teachers may use one or more of the following approaches: didactic large or small group activities, interactive large or small group teacher directed activities, interactive small group activities monitored by the teacher, and/or individual work.

## Didactic Approaches

Individual FM systems were originally designed in the early seventies when **didactic approaches** focussed on teacher-student interaction (rather than student-student interaction) were the standard in education. Thus, the concept of a single transmitter worn by the teacher sending a one-way speech signal to the students was appropriate. As teaching practices began to focus more on activity-based experiential learning where peer interaction was a primary vehicle for language learning, traditional FM system design did not change. The ability to amplify peers' voices is still limited. Therefore, the ideal application for FM usage continues to be classrooms where didactic teaching is the norm.

## Interactive Activities

To maximize the benefit of sound-field FM systems in **interactive large or small group teacher-directed activities**, the use of a pass-around microphone patched into the teacher transmitter to amplify students' voices in discussion is ideal and an excellent way to teach speaker and listener strategies. The presence of the microphone clearly identifies the speaker, and passing the microphone from one student to the next encourages an orderly discussion. The teacher, however, must be prepared for a slower pace of discussion to accommodate physically passing the microphone to the student speaking. Where the pass around microphone option is not available, passing the teacher transmitter from child to child during discussion is desirable. This is only viable, however, in primary or junior grades where the children are grouped around the teacher in close proximity. It is not feasible to pass the teacher transmitter from child to child in the row seating arrangement seen in the older grades. The other alternative, of course, is to have the teacher repeat the student's answers so that they are amplified for the rest of the class.

## Small Group Activities

There are several considerations in use of the FM system during **small group activities** monitored by the teacher. Teachers can use the transmitter and carry a portable speaker to that location, or turn on the speaker in the vicinity of the small group when they are talking to the group, in order to amplify their voices. Amplifying the children's voices for each other when the teacher is not present is problematic though, and peer discussion is the primary avenue for learning in many classrooms. If the sound-field FM system has been placed in the classroom specifically for a child with hearing loss, the teacher may wish to leave his/ her transmitter with the small group in which the particular child has been placed, while ensuring that other speakers in a multiple speaker system are turned off. Of course, the teacher is then unable to present an amplified signal to any of the other small groups. It is necessary to discuss with teachers how they interact within small group activities; often the amount of commentary by the teacher visiting each of the small groups is limited. It is possible that the audiologist and the teacher may decide not to amplify speech during these activities.

## Individual Work

By the time that children are in the third grade, teachers may assign projects requiring **independent efforts** to research, design, and write. It is important to determine how much time is spent in these kinds of activities. In one classroom visited by the author, 90 percent of classroom time was dedicated to individual student projects. Use of a sound-field FM system in this classroom would not be warranted.

## Implications for Installation

It is helpful to sit down with the teacher and draw up a schedule of the type of activities conducted in the classroom and the length of time spent in each type of activity, on a form similar to the one included in Appendix A. This information will help to determine if a sound-field FM system can be effectively used, the type of speaker arrangement suited to the learning style of the classroom, the hardware modifications or accessories necessary, if any, and recommendations for implementation of the system during daily activities.

## TEACHER VARIABLES

Successful introduction of sound-field FM systems is dependent on teacher receptivity to the concept of amplification and to followup provided by support personnel. Teachers are often encouraged by the reports of improved listening and an increased rate of instruction attributed to less need for repetition seen with students in classes using sound-field FM systems. However, typically it is not "talking about" sound-field FM systems that persuades classroom teachers to experiment. Generally, demonstration and actual usage are the factors that provide convincing evidence. Although changes in student's behaviors take some time to observe, the most obvious change in the first few minutes of use is the reduction in the teacher's vocal intensity. Depending on the subject area taught, the background noise levels in the class, and individual personalities, some teachers are at higher risk for vocal fatigue than are others. Teachers such as French instructors in the primary grades who provide only oral instruction all day long have reported to the author that they often contract laryngitis a few weeks after school begins. Physical education teachers can easily abuse their voices when attempting to project in highly reverberant and/or noisy gymnasiums. Access to a demonstration system is the most effective way to show teachers the benefits of enhancing signal for student listening and reduction of teacher vocal fatigue (see Chapter 12 for further information on demonstration systems).

A teacher's resistance, if any, to use of sound-field FM amplification is the use of the hardware, and the reluctance to be "on the air" throughout the day. Once installed, some teachers may not use the system as much as desired during the day. They comment that the children do not "really need" the amplification during particular activities. The ease of use of the transmitter overcomes the resistance of most teachers to the hardware. The audiologist's job is to make the speaker choices and placements simple and practical for the teacher to implement. Unlike personal FM systems, remembering to turn the transmitter off during conferences with individual children is an easy matter with sound-field systems since the teacher is always aware of the amplified signal within the room. There have been cases reported though of teachers going to the staff room and forgetting to turn their transmitter off. Students gathered around the amplified classroom with great glee to listen to the private discussions of staff members.

Inservice training with all teaching staff prior to using the sound-field system is important to overcome the attitude that students can hear adequately in the classroom without amplification. Because students have lived for many years with a reduced signal in a poor acoustic environment, it is assumed by many teachers that students cope satisfactorily. Yet the fatigue that children develop in a difficult listening environment has received little attention to date from audiological researchers. It is reasonable to assume that children in poor acoustic environments must expend considerable energy simply extracting the speech signal. If the speech signal is optimized using sound-field FM technology and additional teaching strategies in the classroom, children can direct their finite amount of energy to processing and comprehending the speech signal, thus capitalizing on what the teacher offers educationally. Therefore, the message that must be conveyed to teachers during inservice training is that although children cope with difficult listening environments, amplification of classrooms enhances children's learning potential.

The most effective inservice tool used by the author to demonstrate the effect of reduced hearing loss on perception of speech is the simulation of hearing loss using foam earplugs (see suggested outline for a hearing loss simulation in Appendix B). This experience gives teachers personal insight into the strains of listening under poor acoustic conditions. The increased ease of listening offered by sound-field FM systems is also strongly supported in anecdotal reports, so having other teachers who already use the system discuss their experiences during the inservice training is encouraging to new users. When all has been said or done, the author has not found one teacher or class that has wished to discontinue using a sound-field FM system once installed with comprehensive inservice training and followup services.

Installation of the system is only the first step in enhancing listening skills in children. **Followup** is essential to sustain the focus on listening skills in the classroom. Visiting the classroom on a monitoring basis once a month maintains the visibility of the importance of hearing in general, and the sound-field FM system specifically. Visitation also permits ongoing discussion of any issues arising from FM usage. Once the teacher is comfortable using the transmitter, encourage the children to participate in the use of the system. Once the audiologist demonstrates good microphone technique to the students, they can practice passing the transmitter from child to child during discussions. This not only enhances perception of the children's voices but also teaches children good speaker-listener strategies that the teacher can carry over to all other classroom activities.

## SUMMARY

Selection, installation, inservice training, and follow-up of sound-field FM systems is a detailed process involving the collaboration of the audiologist, the teacher, and the children in the classroom. When the variables affecting the learning environment are understood, systems can be selected and used appropriately, and the benefits for children's learning are substantive. Key points from this chapter are summarized in Table 7–1.

**Table 7–1.** Key points from this chapter.

| Key Points |
| --- |
| • Understanding of the classroom teaching style and learning style will determine the appropriateness of the sound-field FM system for a classroom. |
| • It is important to ensure that amplified sound will not interfere with adjacent groups or classes in open concept areas and split-grade classrooms. |
| • Selection of types of speakers and location of speakers is dependent on the layout and teaching style used in the classroom. |
| • It is essential to meet with the classroom teacher prior to installation of a sound-field FM system to determine the most appropriate type of sound-field system and the layout of speakers in the room. |
| • Knowledge of the learning style used in the classroom can assist in selection of additional accessories such as the pass around microphone. |
| • Inservice training with school staff when the sound-field FM system is introduced and ongoing monitoring by support personnel is essential to maximizing the benefit of amplification. |

# APPENDIX A
# SAMPLE FORM TO DETERMINE
# CLASSROOM REQUIREMENTS

| Type of Activity | Length of Time | Teacher Involvement | Use of Sound-Field FM and Speaker Placement |
|---|---|---|---|
| Morning Schedule | | | |
| | | | |
| | | | |
| | | | |
| | | | |
| Afternoon Schedule | | | |
| | | | |
| | | | |
| | | | |

Recommendations:

Use of sound-field FM system:    Yes___No___
Type of speaker arrangement:    Portable___Fixed____
Number of speakers:    _____
Modifications required:    Yes___No___
                          describe_____
Accessories required (list):    _____

# APPENDIX B
# SIMULATION OF HEARING LOSS:
# AN INSERVICE TRAINING TOOL

Direct experience often produces optimum learning. Most teachers who are faced with the prospect of a child with hearing loss in their classroom for the first time, express concern about their ability to address the child's needs in their class. By giving the teacher some direct experience with hearing loss, you can provide them with

- an empathetic understanding of the communication demands on the child with hearing loss in the classroom.
- an understanding of the teaching strategies that are detrimental to communication in the classroom.
- an understanding of the teaching strategies that are beneficial to the child with hearing loss in the classroom.

Use of foam earplugs can simulate a mild conductive hearing loss of approximately 25 to 35 dB. The following points are important to emphasize to school staff.

- The simulation only creates a mild hearing loss, and so students with moderate, severe, or profound hearing loss will experience greater difficulty than that experienced with the earplugs.
- The simulation reflects what children with mild hearing loss may hear without a hearing aid, or what children with moderate or moderately severe hearing loss may hear with the hearing aid on.
- Use of the earplugs simulates a conductive rather than a sensorineural hearing loss, since the earplugs are simply impeding the passage of sound through the external ear. This is an important distinction, since the staff must realize that the distortion of speech sounds and the susceptibility to noise seen with children with sensorineural hearing loss cannot be simulated through the use of earplugs alone.
- The simulation produces an accurate perception of the hearing loss often seen with children with recurrent otitis media. Although many teachers may not have experiences with children with sensorineural hearing loss, all primary teachers will have a number of children in their classes each year with histories of recurrent otitis media. (Otitis media is the single

most common reason for a child to visit the family physician, and the most common cause of hearing loss in children.)

## Suggested Procedure

After explaining the purpose of the exercise, hand out a pair of foam earplugs to each group member. Ask the participants to hold the plugs by the rounded edge and roll them between their fingers to compress them to approximately 1/3 to 1/4 of their original size.

Then have everyone insert the compressed plugs into their ear canals so that the canals are completely occluded. If the participants do not hear a clear difference in the loudness of the sound after inserting the plugs, the plugs have not been inserted correctly. Have the individuals remove and reinsert the earplugs. Then ask the participants to get out a sheet of paper and pencil to write down what you say. *There are a number of concepts that you want to demonstrate during the simulation.*

- *The farther away the speaker is from the listener, the more difficult the listening task.*
- *Restricting speechreading cues makes the listening task more difficult.*
- *Presence of background noise increases the difficulty of the listening task.*
- *The type of material presented will vary the difficulty of the task. Single words are much more difficult to identify than is sentence material, where contextual clues can provide a great deal of information.*
- *The intensity of vowels is greater than that of consonants, thus increasing the ease of vowel recognition.*
- *High frequency consononants such as /s/, /ʃ/, /ch /, /k/, /t/, and the voiceless /th/ are usually the most difficult sounds to hear, particularly the /ʃ/ and voiceless /th/, since they are the softest of all of the consonants.*
- *Listening under difficult conditions is fatiguing, resulting in a tendency to tune out or daydream.*
- *Listening can be very frustrating when speakers are far away, or are covering their mouth, or when background noise is present. The listener may experience anger or frustration towards the speaker or towards the sources of background noise.*
- *Additional visual supplements such as writing on the blackboard or the overhead projector can be of great assistance in following the conversation, and reduce the strain of listening.*

Ask participants to write numbers 1 to 15 on the side of the page. In order to demonstrate the above concepts, present words and sentences in the following way.

Write the word....

1. please  BY HEARING ALONE
2. great   (MOUTH COVERED);
3. sled    QUIET CONVERSATIONAL LEVEL;
4. pants   MOVE AROUND WHILE YOU ARE
5. rat     TALKING

Write the word.....

6. bad     BY HEARING ALONE
7. pinch   (MOUTH COVERED);
8. such    CREATE BACKGROUND NOISE
9. bus     (PAPERS RUSTLING, KEYS JINGLING,
10. need   BOOK DROPPING ON FLOOR.....);
           QUIET CONVERSATIONAL LEVEL;
           MOVE AROUND WHILE YOU ARE
           TALKING

Write the word....

11. ways   BY HEARING AND SPEECHREADING
12. five   (MOUTH UNCOVERED);
13. mouth  QUIET CONVERSATIONAL LEVEL;
14. rag    BACKGROUND NOISE SPORADIC
15. put

Now ask the participants to number their page from 1 to 10 and tell them that you will now say some sentences.

1. Walking is my favorite exercise.
2. Here's a nice quiet place to rest.
3. Somebody cleans the floors every night.
4. It would be much easier if everyone would help.
5. Open your window before you go to bed.
        BY HEARING ALONE
        (MOUTH COVERED);
        BACKGROUND NOISE SPORADIC

6. Do you think that she should stay out so late?
7. How do you feel about beginning work at a different time every day?
8. Move out of the way.
9. The water is too cold for swimming.
10. Why should I get up so early in the morning?
    BY HEARING AND SPEECHREADING
    (MOUTH UNCOVERED);
    BACKGROUND NOISE SPORADIC

.................................................................................

- It is important to use a quiet conversational voice level rather than a normal conversational level for maximum effect.
- Because sentences are considerably easier to identify than are single words, they are presented through hearing alone in noise, rather than in quiet.
- The background noise can be sporadic or continuous; the listeners will experience the frustration in either situation.
- When moving around, ensure that you rotate around the entire room so that everyone can experience both optimal and least desirable listening conditions.

.................................................................................

Then have the participants take up their answers WITH THE EARPLUGS STILL INSERTED. When a person gives his or her answer, ensure that the rest of the group has heard it. If not, ask the person to change the way that he or she has presented the answer so that others will understand better (such as repeating the response, saying the word or sentence louder, facing the group, spelling the word, or adding an accompanying gesture). Write down the various answers on a chartboard or overhead to provide a visual supplement. Underline the correct answer from all of the choices provided by the participants.

Once all of the words and sentences have been reviewed, HAVE THE GROUP TAKE OUT THE EARPLUGS. Initiate a group discussion of the following issues:

- their emotional reactions to the overall experience.
- the causes of specific frustrations experienced.
- insights about the experiences of children with hearing loss in the classroom.
- ways in which they could change their teaching strategies to address the needs of children with hearing loss.

The discussion deepens the experience of the simulation of hearing loss and allows participants themselves to determine the necessary changes in teaching strategies.

# CHAPTER
# 8

# SOUND-FIELD AMPLIFICATION: A REVIEW OF THE LITERATURE

*Gail Rosenberg*
*Patricia Blake-Rahter*

For nearly 20 years manufacturers have produced sound-field FM amplification systems. However, until definitive research became available to support the use of this technology in classrooms, interest in these systems was not overwhelming. In the past two years, the market has become competitive with a variety of wireless and wired sound-field FM and VHF amplification devices now available. New options are emerging quickly to meet the demands of school districts seeking to improve listening environments. This chapter contains a review of relevant research that chronicles the history of sound-field FM amplification studies and also demonstrates the validity of this listening enhancement technology. Some researchers have concentrated on special populations and others have sought to evaluate the effectiveness of sound-field FM amplification in regular education classrooms. **The efficacy of sound-field FM amplification is supported by results of research studies that have demonstrated changes in students' academic achievement, speech recognition scores, attending skills, and learning behaviors.** Table 8–1 provides a summary of sound-field FM amplification studies.

## TEACHER RESPONSE TO SOUND-FIELD FM AMPLIFICATION

Teacher response to sound-field FM amplification in the classroom has been overwhelming once they have had opportunity to

**Table 8–1.** Summary of sound-field amplification research.

| Investigators | Student Population | Sound-field Amplification Results |
|---|---|---|
| Sarff (1981); Ray, Sarff & Glassford (1984) | MARRS project (4th–6th grade students with minimal hearing loss, academic deficit, & normal learning potential. | Greater academic achievement at a faster rate & at 1/10th the cost of instruction in unamplified resource rooms. |
| Crandell & Bess (1986) | 20 students with normal hearing in typical classrooms (S/N = +6 dB, RT = 0.6 sec) | Significant improvement in sentence recognition ability under amplified condition. |
| Berg, Bateman & Viehweg (1989) | Regular education junior high school students | Students & teachers preferred use of sound-field amplification, improved student listening & understanding, & ease of listening & teaching. |
| Blair, Myrup & Viehweg (1989) | 10 students (CA: 7–14 yrs.) with mild-moderate SNHL | Average 12% improvement in word recognition score compared to personal hearing aids alone. |
| Flexer (1989, 1992), Osborn, VonderEmbse & Graves (1989) | MARCS project (K–3rd graders in regular education classes) | Greater improvement on *Iowa TBS* for younger students & least academic growth for students who failed hearing screenings in umamplified classes. |
| Gilman & Danzer (1989) | 9 amplified & 9 control classes for 2nd & 4th grade regular education students | Reduced teacher voice fatigue, increased student attentiveness, & ability to hear classroom instruction. |
| Jones, Berg & Viehweg (1989) | Kindergarten students with normal hearing (n=18) and mild hearing loss (n=18) | Higher listening scores for both groups under close listening at 4 ft. & with sound-field amplification. |
| Allen & Patton (1990) | 1st & 2nd grade students with normal hearing | Significant 17% increase in overall student on-task behavior. |
| Benafield (1990) | 4 & 5 year old preschoolers with speech-language delay | More appropriate comments & some improvement in attending behaviors. |
| Flexer, Millin, and Brown (1990) | Primary age children with developmental disabilities | Significant improvement in word recognition scores. |
| Neuss, Blair & Viehweg (1991) | Students with minimal hearing loss | Improved word recognition scores in noise. |

*(continued)*

**Table 8–1.** *(continued)*

| Investigators | Student Population | Sound-field Amplification Results |
|---|---|---|
| Schermer (1991) | First grade students with normal hearing and minimal hearing loss | Higher standardized reading test scores especially for minimal hearing loss students. |
| Ray (1992) | MARRS validation (4th–6th graders with minimal hearing loss & academic deficit) | Academic performance at or above average & students in unamplified resource rooms performed at 0.4 SD below average. |
| Flexer, Richards & Buie (1993) | 283 first grade students with & without known histories of hearing problems | Higher SIFTER scores for "at risk" students; lowest scores for"at risk" unamplified students. |
| Zabel & Tabor (1993) | 145 regular education 3rd–5th grade students | Improved spelling test scores at close and distant seating positions. |
| Crandell (1993) | 20 students with normal hearing | Significantly higher word recognition scores at 12 & 24 feet. |
| Rosenberg, Blake-Rahter, Allen & Redmond (1994) | ICA project (855 K–2nd grade students in regular education classes) | Significantly higher scores for listening & academic behaviors & academic skills; greatest gains for amplified Kg. |
| Crandell (1994) | 20 non-native English children | Improved perception at distances of 12 and 24 feet. |

use this equipment even on a limited basis. A teacher survey conducted by Allen (1993) showed that teachers who have experience using sound-field FM systems in their classrooms value it over eight other types of instructional delivery equipment such as overhead projectors, televisions, computers, VCRs, CDROMs, and filmstrip projectors. When a group of 30 teachers with at least three days experience using sound-field FM amplification systems rated their first choice of instructional delivery equipment, sound-field FM amplification systems received 34 percent of the top choices, followed by overhead projectors (18 percent), and computers (16 percent). Interestingly, two additional groups of teachers, one with indirect exposure to sound-field FM amplification and the other with no exposure to this equipment, rated the overhead projector as their top choice.

## SOUND-FIELD FM AMPLIFICATION SYSTEM OPTIONS

Equipment options available for sound-field FM amplification systems are increasing and can be bewildering. To date, only two researchers have investigated the efficacy of sound-field FM amplification systems. Mills (1991) and Crandell (1993) have conducted comparative studies on commercially available systems. Mills (1991) compared three systems (OMNI 2001 Wireless Classroom Amplification System, Lifeline Freefield Classroom Amplification System, Phonic Ear Easy Listener Freefield Sound System) that were used for a two-month period with at-risk preschool and developmental kindergarten students. Teacher appraisal was used to determine which system was the most cost-effective and maintenance free while providing high sound fidelity. None of the three systems was without some type of problem; however, the Phonic Ear Easy Listener Freefield system proved to be the most trouble-free, easiest to operate, and moderately priced.

Crandell (1993) focused on comparison of speech-recognition abilities of 20 children with normal hearing when using four available FM systems (Comtek OMNI 2001, Lifeline Freefield Classroom Amplification System, Phonic Ear Easy Listener Freefield System, and Radio Shack). A 3-speaker paradigm was used in a typical acoustical classroom environment. Acoustical analyses revealed relatively comparable frequency responses of the four systems. Speech recognition results showed that while each system improved perception scores, there were differences among the systems. Specifically, the Radio Shack and Lifeline systems produced significantly higher speech recognition scores than did either of the other two sound-field FM amplification systems.

## ACADEMIC ACHIEVEMENT

The benchmark 3-year longitudinal project called Mainstream Amplification Resource Room Study (MARRS) was conducted in the Wabash and Ohio Valley schools in southern Illinois from 1977–1980 (Sarff, 1981). The MARRS project compared academic progress of students in grades four through six who received treatment under different amplification conditions. These groups of students were examined for: (1) minimal-to-mild hearing loss (pure-tone thresholds between 10 and 40 dB HL); (2) co-existing learning deficit, and (3) normal learning potential. Approximately half of the children remained in regular classrooms where sound-field FM amplification systems were in use, the remaining children received traditional instruction in unamplified rooms. The sound-field FM amplification system utilized a 2-speaker arrangement and was in use an average of three hours per day. Both groups showed gains in reading, language arts, and total composite scores based on Scientific Research Associates (SRA) Achievement Series test data. However, the greatest improvement was documented for the group receiving instruction in amplified classrooms. Not only did these students show significant gains in academic achievement, but they also were noted to achieve in reading and language arts at a faster rate, to a higher level, and at one-tenth the cost of students taken from regular classes and provided instruction in a resource room setting. It is also important to note that the younger students (fourth and fifth graders) showed the greatest degree of academic growth. The MARRS project achieved national validation status in 1981 as part of the National Diffusion Network of the U.S. Department of Education (Ray, 1987; Ray, Sarff, & Glassford, 1984; Sarff, 1981).

Following the landmark MARRS study, the use of FM sound-field amplification was introduced as an innovative prevention and intervention strategy in classrooms for students with normal hearing as well as students judged to be at risk for listening and learning problems. Project MARRS continues to function as part of the National Diffusion Network and offers assistance to adoption sites. Adoption data validated in 1992 supported previous findings that students with minimal hearing loss who are instructed in unamplified classrooms perform academically at an average level approximately 0.4 standard deviation (SD) below normal. Students with minimal hearing loss in amplified classrooms were found to perform at or above average (Ray, 1992).

Over a three-year period, the Putnam County Office of Education investigated the efficiency and cost-effectiveness of sound-field amplification for improving the quality and consistency of oral instruction for lower elementary children in nine rural school districts in Ohio (Flexer, 1989, 1992; Osborn, VonderEmbse & Graves, 1989). Project MARCS (Mainstream Amplification Regular Classroom Study) utilized the OMNI 2001 sound-field FM amplification system with a 2-speaker paradigm in 17 kindergarten through third grade classrooms matched with 17 unamplified classrooms. Results of the Iowa Test of Basic Skills (TBS) indicated higher achievement levels for students in the experimental classrooms in the following subtest areas by grade level for the first year: listening and language (kindergarten and first grade), vocabulary (first grade), math concepts (second and third grade), and math computation (third grade). For the second year, significant findings were reported for experimental classrooms in three of the four grades: word analysis (kindergarten and first grade), math concepts (first and third grade), math problem solving (first grade), and math computation (third grade). A general trend showed that the younger the student, the greater the difference between the control and experimental group's achievement test scores. Those students who failed to pass pure-tone and tympanometry screening and were placed in unamplified classes showed the poorest overall performance on the TBS.

Classroom observations were also a part of Project MARCS. Experimental findings indicated that use of sound-field FM amplification encouraged teacher mobility, increased the number of students participating, produced a more consistent teacher rate of speech, and showed better student transition between classroom activities and general activity levels. There were fewer special education referrals and learning disabilities placements in the schools with the greatest number of sound-field FM amplification systems in the lower elementary grades. Informal MARCS evaluation by teachers in experimental classrooms generally indicated more consistent student attending skills, a reduction in teacher's vocal strain and voice fatigue, and increased versatility in instructional techniques. At the end of three years, 46 classrooms were using sound-field FM amplification which suggests that these educators were convinced that it was helping students listen and learn better.

Schermer (1991) conducted a study of reading achievement in first grade students. Results showed higher achievement on the Gates-MacGinitie Reading Tests for students in amplified classrooms, particularly for students with primarily mild, fluctuating, conductive hearing losses. Students with known mild hearing loss who were not placed in amplified classrooms showed a decrease on post-test scores. A recommendation of this study was to explore the effectiveness of sound-field FM amplification as an alternative to special education personnel required for resource room instruction. We now know that sound-field FM amplification is a viable alternative that will meet placement requirements for many students with known hearing loss. Subsequently, this allows other students in the class to benefit from the enhanced access to auditory information.

# SPEECH RECOGNITION

Crandell and Bess (1986) were the first to report findings of improved speech-recognition scores when students used sound-field FM amplification. These authors examined its benefits for 20 students with normal hearing in a typical classroom setting (signal-to-noise ratio [S/N] of +6 dB; reverberation time [RT] of 0.6 second at 6, 12, and 24 feet). Research findings showed significant improvement in students' ability to recognize sentence information under the amplified condition at 12 and 24 feet from a speaker.

Blair, Myrup, and Viehweg (1989) compared word recognition abilities of 10 students (7–14 years of age) with mild-to-moderate sensorineural hearing loss under three conditions: (1) personal hearing aids in conjunction with a 2-speaker sound-field system, (2) personal hearing aids coupled via miniloop to a personal FM system, and (3) personal hearing aids only. These investigators used a 2-speaker ceiling-mounted presentation to achieve sound-field FM amplification. Use of personal hearing aids in conjunction with the sound-field FM amplification system yielded an average improvement of 12 percent in word recognition scores compared to those obtained when students used only personal hearing aids. However, the greatest gain in word recognition scores was achieved when students used their hearing aids coupled with personal FM systems; this resulted in an additional 5 percent improvement over the sound-field FM amplification condition. These findings clearly support use of a sound-field FM amplification system as an additional tool for access to auditory information for students with properly functioning hearing aids. The investigation did not examine use of the personal FM system in conjunction with the sound-field FM amplification system, which is a common practice in schools today.

The effects of speaker-listener distance and sound-field FM amplification on the speech perception of 18 normal-hearing and 18 kindergarten students with mild sensorineural hearing loss were studied by Jones, Berg, and Viehweg (1989). Students were given a speech recognition test under three conditions: (1) 12 feet from the tape player, (2) 4 feet from the tape player, and (3) sound-field FM amplification via ceiling-mounted speakers. Results revealed significantly higher perception scores for both groups under the close (4 feet) and sound-field FM amplification listening conditions compared to distant listening at 12 feet. Although the average sound-field score was slightly higher than the average close listening score, significant differences were not seen. This study demonstrated the advantage of improving the S/N ratio in kindergarten classrooms either by physically decreasing student-teacher distance or by using sound-field FM amplification.

Flexer, Millin, and Brown (1990) examined the effects of sound-field FM amplification on word identification by primary-age children with developmental disabilities. An SRT-100 Classroom Acoustic Sound Field System was used with a 2-speaker design. Audiological evaluation data indicated that only one of the nine students had normal hearing sensitivity and/or normal immittance results. Six students had histories of persistent hearing loss and had been referred repeatedly for medical treatment. Results indicated significant improvement in word recognition scores for the developmentally disabled youngsters under the amplified listening condition. In the unamplified condition, students made almost three times as many errors as in the amplified listening condition.

Neuss, Blair, and Viehweg (1991) investigated the word recognition abilities of students with minimal hearing loss under three listening conditions: (1) personal hearing aid only, (2) unaided, and (3) sound-field FM amplification only. Students had pure-tone averages (PTAs) between 15 and 40 dB HL and were hearing aid users. A Realistic sound-field FM amplification system with a 3-speaker arrangement was used to provide amplification in this study. Monosyllabic word recognition scores obtained in noise showed significant improvement when students used sound-field FM amplification, but not when personal hearing aids were used.

The performance of 145 normal-hearing third, fourth, and fifth graders on spelling tests under unamplified and amplified conditions was reported by Zabel and Tabor (1993). The unamplified listening condition was presented at 0 dB S/N (typical classroom) and at +12 dB S/N ratio in the amplified condition. The Lifeline system with a 4-speaker arrangement was used during amplified testing. The investigators used tape recorded Curriculum-Based Measurement spelling tests and found that all classes achieved significant improvement in spelling scores under amplified conditions.

Finally, Crandell (1993) evaluated commercially available sound-field FM amplification systems as a function of student's word recognition scores at various speaker-listener distances of 6, 12, and 24 feet. Findings showed that speech-recognition scores were enhanced, compared to unamplified listening, using each of the FM systems. In addition, students achieved elevated word recognition scores in the amplified treatment at speaker-listener distances of 12 and 24 feet.

# ATTENDING SKILLS STUDIES

Although most of the sound-field FM amplification studies have been with elementary-age children, Berg, Bateman, and Viehweg (1989) conducted a study in junior high school students. The effects of sound-field FM amplification on attention, understanding, ease of listening, ease of teaching, and preference for the amplified listening condition were rated by students and teachers. Results showed that both groups preferred the amplified setting and that the FM equipment improved student attention and understanding, as well as ease of listening and teaching. The OMNI 2001 and Realistic sound-field FM amplification systems were used in this project. Interestingly, student and teacher ratings were not significantly influenced by either system.

Gilman and Danzer (1989) conducted a comprehensive math, reading, and language pre- and post-assessment of second and fourth grade students in nine amplified and nine unamplified classrooms. The OMNI 2001 system was used in this project. Findings showed that sound-field FM amplification reduced teacher voice fatigue, increased student attentiveness to verbal instruction and activities, and increased students' ability to hear classroom instruction.

The effects of sound-field FM amplification on a group of randomly selected, normal-hearing first and second grade students' on-task behavior were studied by Allen and Patton (1990). Using a systematic observation protocol they found that under the amplified condition, students were more attentive, less distractible, and required fewer repetitions by the teacher. In the amplified condition a significant increase of 17 percent was documented for students' overall on-task behavior.

By 1990, sound-field FM amplification studies of special populations other than students with known hearing loss began to emerge in the literature. The effects of amplification on attending behaviors of four- and five-year-old speech and language-delayed preschoolers were reported by Benafield (1990). The Phonic Ear Easy Listener Free Field system was installed in a language enrichment preschool classroom. Although there was no significant effect on subjects' verbalizations when using the system, there was a trend toward greater appropriate subject comments during amplified instruction. In addition, preschoolers showed an increase in several physical attending behaviors during times when sound-field FM amplification was in use.

## LEARNING BEHAVIORS

Young children with early and continuing histories of hearing problems are potentially at risk for learning. Flexer, Richards, and Buie (1993) examined teacher-perceived performance of two groups of first grade children (n=283) under amplified and unamplified listening conditions. Teachers were asked to note differences between the group of "at risk" children (early and continuing histories of hearing problems) from a "no risk" group (without known histories of hearing problems). The Screening Instrument for Targeting Educational Risk (SIFTER) (Anderson, 1989), was completed by teachers at four intervals during the school year. The SIFTER includes 15 student behavior statements in the areas of academics, attention, communication, class participation, and school behavior. SIFTER. data showed better overall ratings for both the "at risk" and "no risk" groups in classrooms with sound-field FM amplification. Students receiving the lowest teacher appraisal were those "at risk" in unamplified classrooms.

## IMPROVING CLASSROOM ACOUSTICS (ICA) PILOT PROJECT

The Improving Classroom Acoustics (ICA) pilot project conducted in four Florida school districts was designed to examine improvement in listening behaviors, academic/pre-academic behaviors, and academic/pre-academic skills for kindergartners, first, and second graders (Rosenberg, Blake-Rahter, Allen, and Redmond, 1994). This special project was funded by the Florida Department of Education, Division of Public Schools, Bureau of Education for Exceptional Students, through federal assistance under the Individuals with Disabilities Education Act (IDEA), Part B. A 12-week observation period was initiated in the fall of 1993 and involved 855 students in 20 experimental (amplified) and 20 control classes. Sound-field FM amplification was provided by the Phonic Ear Easy Listener Free Field system with a 4-speaker arrangement. Teachers participated in comprehensive inservice training using materials developed for the project (Florida Department of Education, 1994). Audiologists conducted inservice training prior to implementation of the project and provided on-site assistance throughout the study.

Ambient noise levels for unoccupied classrooms averaged at 50.25 dB(A) while occupied classrooms averaged at 64.05 dB(A). Teachers in experimental classrooms showed an average gain of +7.52 dB(A) when using the sound-field FM amplification system.

All students were observed pre-, mid- (6 weeks), and post-treatment (12 weeks) using the Listening and Learning Observation (LLO) form devised for the project. Ten randomly selected students in each class were also observed using an adaptation of the Evaluation of Classroom Listening Behaviors (ECLB) (VanDyke, 1985). Hearing screening results showed pass rates of 80.70 percent at 15 dB HL, 99.50 percent at 20 dB HL, and 93.45 percent for tympanometry.

Results of the pilot project showed that students in amplified classrooms (n=430) demonstrated a significantly greater change in listening behaviors, academic/pre-academic behaviors, and academic/pre-academic skills, and at a faster rate than their peers in unamplified classrooms (Figure 8–1). Although both groups showed equivalent scores for academic/pre-academic skills at pre-treatment, the experimental group showed significantly greater gains based on teacher appraisal after 6 weeks of sound-field FM amplification. Results of paired t-tests showed significant improvement for the experimental group with the greatest improvement noted at mid-treatment. While both groups showed significant change from pre- to post-treatment, the mean differences for the experimental group were notably higher (Table 8–2). Kindergarten students in amplified classrooms achieved the greatest gains in each of the three aforementioned areas over the 12-week observation. Similar results were obtained for 384 students who were also observed using the ECLB.

Project evaluation showed overwhelming support from students, teachers, parents, and administrators toward use of sound-field FM amplification. When asked if they wanted to use the listening equipment again next year, 93.3 percent of the students responded affirmatively. Teachers used the systems an average of 4.38 hours per day and indicated decreased vocal strain as the primary benefit from using the sound-field FM amplification system. All of the teachers agreed that they enjoyed using the FM system, felt prepared to use the equipment following inservice training, found it easy to use, and wanted to keep it in their classroom.

Slightly more than 20 percent of the parents who completed the evaluation form (n=279) observed the FM system in operation. Parents gave their highest rating to the item requesting that their child be able to use the sound-field FM amplification system again next year. Administrators acknowledged that class instruction and management seemed to be enhanced when teachers used classroom amplification and also commented that it seemed to

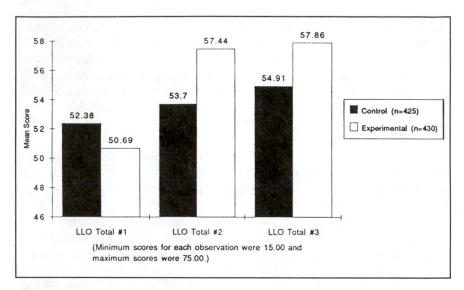

**Figure 8–1.** Listening and Learning Observation (LLO) instrument mean total scores at pre-treatment (#1), mid-treatment (#2), and posttreatment (#3) for control and experimental groups. Observations occurred at 6-week intervals.

**Table 8–2.** Paired t-test results (mean differences and P-values) for observations 1–2 (6 weeks) and 1–3 (12 weeks) for Listening and Learning Observation (LLO) total score, LLO Listening Behaviors, LLO Academic/Pre-Academic Behaviors, and LLO Academic/Pre-Academic Skills for control and experimental groups.

| Observation Tool | Treatment | Observation 1–2 | | Observation 1–3 | |
|---|---|---|---|---|---|
| | | Mean Difference | P-value | Mean Difference | P-value |
| LLO Total | Control | −1.32 | NS | −2.53 | <.0001 |
| | Experimental | −6.76 | <.0001 | −7.17 | <.0001 |
| LLO Listening Behaviors | Control | −0.95 | NS | −1.61 | <.0001 |
| | Experimental | −4.61 | <.0001 | −4.86 | <.0001 |
| LLO Academic/ Pre-Academic Behaviors | Control | −0.20 | NS | −0.49 | <.0001 |
| | Experimental | −1.14 | <.0001 | −1.19 | <.0001 |
| LLO Academic/ Pre-Academic Skills | Control | −1.17 | NS | −0.43 | <.0001 |
| | Experimental | −1.01 | <.0001 | −1.13 | <.0001 |

help "save the teacher's voices." Perhaps some of the most convincing information about sound-field FM amplification in the classroom is that gleaned from testimonials given by students, parents, teachers, and administrators. A sampling of key comments from the ICA pilot study are displayed in Table 8–3.

A major concern of teachers, parents, and administrators, however, was the cost of providing these systems. Incidentally, the cost for sound-field FM amplification for the 12-week pilot project (430 students and 20 teachers) was $6.48 per week, or approximately 11 cents per day per person. (See Table 8–4.) Applying a longevity factor over time will serve to further decrease the direct cost for amplifying 20 classrooms. The ICA continuation project should provide data on more than 1200 students as it involved two additional observations on students in the pilot project and included two more districts which conducted another 12-week pilot during the second semester.

## SUMMARY AND CONCLUSIONS

Many of these representative studies found that younger children tended to exhibit greater behavior changes and academic growth than did older children. This may be attributable in part to the fact that younger children are often plagued by such problems as otitis media and require speech to be louder in order to hear more clearly (Crandell & Smaldino, 1992; Flexer, 1992). **Collectively, results of sound-field FM amplification research suggest significant benefits for students in the areas of academic achievement, speech recognition enhancement in quiet and in noise, and on-task behavior.** Students, teachers, parents, and school administrators have indicated approval for this unique FM technology, although cost remains an area of concern. Audiologists must assume a leadership role in promoting the concept and benefits of sound-field FM amplification as a means to help not only students with known hearing loss but as an avenue to allow all students the opportunity to enjoy an enhanced listening and learning environment.

**Table 8–3.** ICA pilot project testimonials from students, teachers, parents, and administrators on the perceived benefits of sound-field FM amplification (Rosenberg, Blake-Rahter, Allen, & Redmond, 1994).

---

**Student Comments**
- The microphone helps me hear better because my teacher has a soft voice and you can't always hear her.
- When my teacher turns off the speakers we cannot hear that good.
- It helps a lot of people who sit in the back and can't hardly hear without the microphone.
- Our teacher lets us read with it and that's good for shy people.
- When other kids tell stories to the class we can all hear real good.
- The speakers make us all pay attention better than we used to last year.
- I hope we get to have it in our classroom next year to hear better some more.

**Teacher Comments**
- Students could hear and understand better.
- The system encourages shy children to speak and share information in front of the class; it builds confidence in all students when the use the microphone.
- It helps students follow directions and enhances listening skills, especially low ability students.
- More seating options are available to students with hearing loss.
- It decreases teacher's voice problems, reduces vocal strain, produces a more relaxed feeling when teaching, and teachers feel less tired at the end of the day.
- It improves student attention; the system assists in getting and holding student's attention.
- It is an effective tool for chalkboard writing, giving tests, and playing tapes in class.
- Everyone benefits by using the FM system in the classroom. Every classroom should have one.

**Parent Comments**
- It improves the student's ability to hear the teacher more easily and clearly.
- It improves the students' ability to understand the teacher more easily.
- It makes learning easier.
- It increases self-confidence when students use the microphone in front of the class.
- It improves the attention and focus of students.
- It improves the students' ability to hear the teacher from any location in the room, when writing on the chalkboard, when speaking softly, and above classroom noise.
- It improves students' ability to concentrate and ignore classroom noise.
- It improves students' behavior.
- It is easier on the teacher when using the FM system.
- Students enjoy using the FM sound-field classroom amplification system. Every classroom should have one.

*(continued)*

---

**Table 8–3.** *(continued)*

**Administrator Comments**
- It saves the teacher's voice and they are less fatigued at the end of the day.
- Students could hear the teacher equally as well from any point in the classroom and able to hear the teacher clearly at all times.
- Students seemed to listen better.
- Students seemed to focus more quickly and consistently.
- Students were more in tune with the teacher.
- Students like using the amplification and felt important.
- Class time was saved because instructions did not have to be repeated.

**Table 8–4.** Cost-effectiveness of sound-field FM amplification equipment for the ICA pilot project.

| | |
|---|---|
| Length of ICA pilot project | 12 weeks |
| Participants | 450 (430 students and 20 teachers) |
| Cost per person per week | $6.48 |
| Approximate cost per person per day | $0.11 |

# REFERENCES

Allen, L. (1993). Promoting the usefulness of classroom amplification. *Educational Audiology Monograph, 3*, 32–34.

Allen, L., & Patton, D. (1990). *Effects of sound-field amplification on students' on-task behavior.* Paper presented at the American Speech, Language, and Hearing Convention, Seattle, WA.

Anderson, K. (1989). *Screening instrument for targeting educational risk (SIFTER).* Austin, TX: PRO-ED.

Benafield, N. (1990). The effects of sound field amplification on the attending behaviors of speech and language-delayed preschool children. Unpublished master's thesis, University of Arkansas at Little Rock.

Berg, F., Bateman, R., & Viehweg, S. (1989). *Sound field FM amplification in junior high school classrooms.* Paper presented at the American Speech-Language-Hearing Association Convention, St. Louis, MO.

Blair, J., Myrup, C. & Viehweg, S. (1989). Comparison of the effectiveness of hard-of-hearing children using three types of amplification. *Educational Audiology Monograph, 1*(1), 48–55.

Crandell, C. (1993). A comparison of commercially-available frequency modulation sound field amplification systems. *Educational Audiology Monograph, 3*, 15–20.

Crandell, C., & Bess, F. (1987). *Sound-field amplification in the classroom setting.* Paper presented at the American Speech-Language-Hearing Association Convention, New Orleans.

Crandell, C., & Smaldino, J. (1992). Sound field amplification in the classroom setting. *American Journal of Audiology, 1*(4), 14–16.

Flexer, C. (1989). Turn on sound: An odyssey of sound field amplification. *Educational Audiology Association Newsletter, 5*(5), 6–7.

Flexer, C. (1992). Classroom public address systems. In M. Ross (Ed.), *FM auditory training systems: Characteristics, selection & use* (pp. 189–209). Timonium, MD: York Press.

Flexer, C., Millin, J., & Brown, L. (1990). Children with developmental disabilities: The effect of sound field amplification on word identification. *Language, Speech and Hearing Services in Schools, 21*, 177–182.

Flexer, C., Richards, C., & Buie, C. (1993). *Soundfield amplification for regular kindergarten and first grade classrooms: A longitudinal study of fluctuating hearing loss and pupil performance.* Paper presented at the American Academy of Audiology Convention, Phoenix, AZ.

Florida Department of Education. (1994). Improving classroom acoustics: Inservice training manual. Tallahassee, FL.

Gilman, L., & Danzer, V. (1989). *Use of FM sound field amplification in regular classrooms.* Paper presented at the American Speech-Language-Hearing Association Convention, St. Louis, MO.

Jones, J., Berg, F., & Viehweg (1989). Listening of kindergarten students under close, distant, and sound field FM amplification conditions. *Educational Audiology Monograph, 1*(1), 56–65.

Leavitt, R., & Flexer, C. (1991). Speech degradation as measured by the rapid speech transmission index (RASTI). *Ear & Hearing, 12*, 115–118.

Lewis, D. (1994). Assistive devices for classroom listening. *American Journal of Audiology, 3*(1), 58–69.

Mills, M. (1991). A practical look at classroom amplification. *Educational Audiology Monograph, 2*(1), 39–42.

Neuss, D., Blair, J., & Viehweg, S. (1991). Sound field amplification: Does it improve word recognition in a background of noise for students with minimal hearing impairments? *Educational Audiology Monograph, 2*(1), 43–52.

Osborn, J., VonderEmbse, D., & Graves, L. (1989). Development of a model program using sound field amplification for prevention of auditory-based learning disabilities. Unpublished Study, Putnam County Office of Education, Ottawa, OH.

Ray, H. (1987). Put a microphone on the teacher: A simple solution for the difficult problems of mild hearing loss. *The Clinical Connection*, Spring, 14–15.

Ray, H. (1988). Mainstream amplification resource room study (MARRS): A national diffusion network project. Wabash & Ohio Valley Special Education District, Norris City, IL.

Ray, H. (1989). Project MARRS—An update. *Educational Audiology Association Newsletter, 5*(5), 4–5.

Ray, H. (1992). Summary of MARRS Adoption Data Validated in 1992. Wabash & Ohio Valley Special Education District, Norris City, IL.

Ray, H., Sarff, L., & Glassford, F. (1984). Soundfield amplification: An innovative educational intervention for mainstreamed learning disabled students. *The Directive Teacher, 6*(2), 18–20.

Rosenberg, G., Blake-Rahter, P., Allen, L., & Redmond, B. (1994). Improving classroom acoustics: A multi-district pilot study on FM classroom amplification. Poster session presented at the American Academy of Audiology annual convention, Richmond, VA.

Sarff, L. (1981). An innovative use of free-field amplification in classrooms. In R. Roeser & M. Downs (Eds.), *Auditory disorders in school children* (pp. 263–272). New York: Thieme-Stratton.

Sarff, L., Ray, H., & Bagwell, C. (1981). Why not amplification in every classroom? *Hearing Aid Journal, 34*(12), 11.

Schermer, D. (1991). Briggs sound amplified classroom study. Unpublished study, Briggs Elementary School, Maquoketa, IA.

VanDyke, L. (1985). Evaluation of classroom listening behaviors. *Rocky Mountain Journal of Communication Disorders.*

Zabel, H., & Tabor, M. (1993). Effects of classroom amplification on spelling performance of elementary school children. *Educational Audiology Monograph, 3*, 5–9.

# CHAPTER
# 9

# CONSIDERATIONS AND STRATEGIES FOR AMPLIFYING THE CLASSROOM

*Carol Flexer*
*Carl Crandell*
*Joseph Smaldino*

Now that the classroom has been made quieter (see Chapter 6), and the teaching style has been identified (see Chapter 7), how do we select sound-field FM amplification systems to fit a particular classroom and learning environment? **How do we reach the goal of providing a 10 dB signal-to-noise (S/N) ratio improvement?**
Several companies have been manufacturing sound-field FM equipment for years, and more are getting into the business as time goes on. The major companies are listed in Appendix A at the end of this chapter. This chapter will identify and discuss issues that need to be considered for the selection, installation, and use of sound-field FM amplification systems. Specifically, these issues on sound-field amplification will be addressed in this chapter: (1) cost factors; (2) appropriate FM carrier frequencies; (3) evaluating the number of available discrete channels used by various manufacturers; (4) number and positioning of loudspeakers; (5) durability, flexibility, portability, and ease of installation of the sound-field amplification equipment; (6) fidelity of equipment; (7) where to position children with a known hearing impairment or perceptual deficit in an amplified classroom; (8) microphone style; (9) inservice training and follow-up issues; and (10) feedback control. Prior to this discussion, however, the reader is directed to Tables 9–1 and 9–2 for advantages/disadvantages of sound-field FM systems.

**Table 9–1.** Advantages of sound-field amplification.

---

(1) The system can be used with each of the populations of "normal hearing" children discussed in Chapter 4, as well as children with SNHL. As indicated in this book, the vast majority of children in our school systems can benefit from sound-field amplification.

(2) In terms of children with mild SNHL, the system can provide benefit while malfunctioning hearing aids or auditory trainers are repaired (see Chapter 4).

(3) Sound-field systems are often the most inexpensive procedure of improving classroom acoustics. Certainly, as discussed in Chapters 1 and 8, such systems are the least expensive of the assistive listening devices used for classroom amplification.

(4) A sound-field system does not stigmatize certain children, which can be the situation with auditory trainers or hearing aids because they require the children to wear hardware.

(5) From 90–95 percent of teachers willingly accept sound-field amplification systems. Moreover, teachers report lessened stress and vocal strain during teaching activities.

(6) Students can utilize the system for oral reports, oral reading, and for asking/answering questions, thus enhancing academic performance.

(7) Sound-field systems can be used to enhance other instructional equipment (e.g., television, cassette tape/CD players).

---

**Table 9–2.** Potential disadvantages of sound-field amplification.

---

(1) Sound-field amplification systems may not provide adequate benefits in excessively noisy or reverberant learning environments. That is, 8–10 dB of amplification may not be enough to overcome the effects of a particularly poor acoustical environment. Specifically, the use of sound-field amplification can increase both the level of the desired direct sound and the undesired reflected sound energy. In addition, because of varying amounts of sound cancellation and distortion, it may be difficult to obtain the desired uniform sound amplification throughout the classroom without proper speaker placement.

(2) If the speaker arrangement is not appropriate for the classroom, the teacher's voice may be amplified excessively for some children, while not providing sufficient amplification for other children. In both situations, the benefits to the student and/or teacher will be diminished.

(3) In smaller classrooms, the amplified sound may be less than 10 dB because of feedback problems associated with the synergistic effects of reflective surfaces and speaker closeness. At present, it is not clear how much benefit limited amplification can provide.

(4) Both the teacher and student need appropriate in-service information and follow-up support for the sound-field system to provide maximum benefit in the classroom.

(5) Sound-field amplification should not be used as a substitute for a personal FM systems for children with moderate-profound hearing impairments.

---

## COST FACTORS

Current commercially available units vary from about $600 to $1450 depending on the quality of the equipment, the number of loudspeakers, service contracts, ease of installation, durability, flexibility, and such special features as extra transmitter microphones for team teaching. Some instructors assemble their own equipment from component parts sold by companies such as Radio Shack. Component parts can be assembled for approximately $250. A high quality, permanently installed public address and sound system such as those used in theaters and public speaking halls can cost thousands of dollars.

Like it or not, school systems are operating on tighter and more restricted budgets. See Chapter 12 for ideas about how to market and obtain funding for FM systems. Cost is mentioned here not only to acknowledge its importance, but also to introduce a caution. It is important to evaluate *ALL* of the issues relative to selecting and installing sound-field FM systems, and not just the cost of the equipment. We certainly want the best quality sound system for the lowest price. However, be aware that a low-fidelity system that is not user-friendly, or that has significant interference, or that often breaks down with no warranty or service contract, can be far worse than not having any sound-field FM equipment at all. A 30-day trial period represents a reasonable time period to evaluate cost-effectiveness.

## RADIO SIGNAL CARRIER FREQUENCY

Different equipment manufacturers may use different frequency bands to transmit the radio signal. Some of the frequency bands, called "business bands," are crowded, and thus may be subject to interference from undesired sources. Much of the component equipment assembled from parts obtained from outlets such as Radio Shack use the 49 MHz band. This frequency band must compete for radio space with walkie-talkies, paging systems, and toys. Consequently, the assembled sound-field FM unit tends to pick up unwanted conversations and distracting interference.

Many manufacturers of less expensive sound-field FM equipment primarily use 174–216 MHz. A newer, favored band is the 900 MHz band, called ultra high frequency (UHF). These bands also can pick up interference, but typically not as much as the 49

MHz band. Frequency bands that are dedicated for low-power FM transmission for individuals with hearing impairment include 72–76 MHz. These frequencies tend to have the least interference and are offered primarily by those manufacturers who also produce personal FM devices.

It is important to find out if the carrier frequency of your prospective sound-field FM unit can be changed by the manufacturer if interference is experienced versus being a unit that is a "throw-away" if it does not work appropriately in your school environment. The purpose of installing sound-field FM technology is to improve the S/N ratio in the classroom thereby facilitating attention and acoustic accessibility of instructional information. Intermittent static, buzzing, clicking, random conversations, or unwanted music transmitted over the loudspeakers is disruptive.

## NUMBER OF DISCRETE CHANNELS WITHIN THE MANUFACTURER'S BAND

Some sound-field FM systems have only 2–4 discrete channels available for use. Therefore, only 2–4 units could be used in the same building without potentially interfering with one another. On the other hand, the more discrete channels that a frequency band is divided into, the poorer the S/N ratio might be. Up to 40 narrowband channels can be accessed using the 72–76 MHz bands. Using those same bands, ten wideband channels can be obtained. To summarize, too few bands limit the number of sound-field units that may be installed in the same building, while too many bands per carrier frequency might cause interference and an unfavorable S/N ratio within the equipment itself (Boothroyd, 1992).

## NUMBER AND POSITIONING OF LOUDSPEAKERS

How many loudspeakers and where to place them is probably the most frequently asked question, and the one that can be answered with the least certainty. We need more data-based studies to investigate these issues because of the many variables present in each individual classroom environment.

Classrooms variables include size, shape, construction, composition, and seating arrangements. Learning styles also vary considerably; whole group versus individual learning, versus small group interactive learning, versus learning centers, versus independent learning, versus self-contained, versus open classrooms. Teaching style is another important variable and includes single-teacher lecture, team teaching, or multiple teaching occurring simultaneously at several learning centers throughout the room. The students themselves add yet additional variables according to their ages, numbers in a class, compliance, and additional disabling conditions they may possess.

Would the same sound-field FM unit loudspeaker arrangement be suitable for the plethora of classroom compositions, teaching styles, and pupils just mentioned? Probably not! To the contrary, amplifying an auditorium or a theater is relatively easy to achieve by following a fixed formula, because the stage is always up front, the actors/speakers are always up front on the stage and the spectators are always seated in stationary chairs in fixed positions throughout the room.

Consequently, we recommend a *pragmatic approach* to ordering and installing sound-field FM equipment. This pragmatic approach considers the individual classroom, the individual teacher and teaching style(s) used in it, and the pupils who learn in that particular classroom.

## Determining the Number of Loudspeakers

Adequate acoustic dispersion of the teacher's voice throughout the classroom is a function of many factors, including the wattage of the sound-field amplifier used, the directionality of the speakers, and the location and number of the speakers in the room. It has been common practice to use multiple speakers in order to accomplish adequate dispersion, although there is little research to base this decision on other than intuition and what sounds good to the installer. It is well recognized that carelessly installed, multiple speakers in a room can alter the acoustic spectrum of the speaker's signal and may produce psychoacoustic effects such as listener fatigue. Single, point speakers may have fewer problems. More research is needed in this area to ascertain the best way to accomplish adequate dispersion in classrooms.

Obviously, the goal is for all students in the learning field to have access to an even, consistent, and favorable S/N ratio at a 10 dB improvement over unamplified speech. Thus, if group learning is the mode of teaching, the entire classroom needs to be amplified evenly; all the pupils need to be able to hear the teacher at all times. The larger the classroom, typically the more loudspeakers (three or four) that need to be used so that each child is closer to a loudspeaker than he or she could be to the teacher. Close proximity avoids loss of critical speech elements as the sound signal is transmitted across a physical space. If angled properly, three or four loudspeakers positioned about 5 feet up on walls, or in a ceiling array could provide "surround" sound for all students.

If the classroom has learning centers, a loudspeaker could be positioned close to each learning center for maximum effective amplification of the critical locations. If only one learning center is used at a time, then the other loudspeakers should be turned off.

If a small resource classroom is used, two loudspeakers could provide an even and consistent S/N ratio throughout the area. In fact, if the room is quite small, and there are only a few students who can be seated close to the teacher, even a single loudspeaker might be effective.

If the classroom and class size is small, with only one teacher-instructed learning center in use at any given time, then a single battery-powered loudspeaker could be carried by the teacher to each teaching location to amplify that specific environment. This single, portable loudspeaker arrangement has worked very effectively in some preschool settings.

For example, Figure 9–1 shows the use of a single, portable loudspeaker to amplify a small learning center. This particular teacher does not use a group instruction paradigm. Rather, all instruction is carried out in small learning centers, and the teacher carries the battery-powered loudspeaker with her to each individual location.

The teacher in Figure 9–2a and b has a primary group learning area in the center of the room, with individual work stations located along the periphery. Therefore, multiple loudspeakers have been arranged to amplify the group learning center.

## Loudspeaker Positioning

Unfortunately, there are no data-based guidelines for loudspeaker placement. **Again, a pragmatic approach is recommended**. It seems logical that sound could be heard most efficiently if it is directed to the ears of the children. Consequently, some manufacturers recommend placement of loudspeakers on speaker stands at the ear level of children. The problem with such placement is that the stands tend to get knocked over by young, active children, or obstructions occur between the loudspeaker and the ears

**Figure 9–1.** A single, portable, battery-powered loudspeaker is carried by the teacher to each small learning center for effective amplification of that learning environment (Courtesy of Audio Enhancement).

of children. Other manufacturers recommend placing loudspeakers at several feet above the seated ears of children with the loudspeakers angled down. For many reasons, the authors have found greater success with this latter recommendation.

The speaker angle of dispersion is certainly an issue. Some teachers, in a practical effort, place loudspeakers on any available bookshelf, facing any which way. Other teachers mount the loudspeakers up high on the walls toward the ceiling. Loudspeakers that are mounted high on walls can cause increased room reverberation if the sound is bounced off the ceiling. The point is that loudspeakers that are not thoughtfully placed can cause **more** reverberation in a room, rather than overcoming a poor acoustic environment. Loudspeakers placed in ceilings also need to be unobstructed and angled to allow a favorable angle of dispersion.

Loudspeakers inadvertently have been turned backwards, had gerbil tanks placed in front of them, or have had books placed in front of them, plus other obstructions. Loudspeakers must remain unobstructed and angled toward the children. Figures 9–3 and 9–4

a

**Figure 9–2.** Multiple loudspeakers are positioned around a central group-learning center to provide an even and consistent 10 dB S/N ratio enhancement throughout that learning environment (Figure 9–2a Courtesy of Audio Enhancement; Figure 9–2b Courtesy of Phonic Ear).

show examples of correct and incorrect loudspeaker positioning. Certainly loudspeakers should never be placed directly across from each other or flush with corner areas.

## Loudspeaker Output Measurement

Refer to Chapter 5 for suggested measurement protocols and worksheets. Measurement strategies include making sound-level measurements, obtaining RASTI scores, and subjective evaluation of the sound of the unit—called "ear cuing." Speech that is heard through a sound-field FM system should be easily and uniformly audible from and at all learning locations in the classroom. In addition, spoken instruction should be comfortably intelligible and nonstressful.

**b**

**Figure 9–2.** *(continued)*

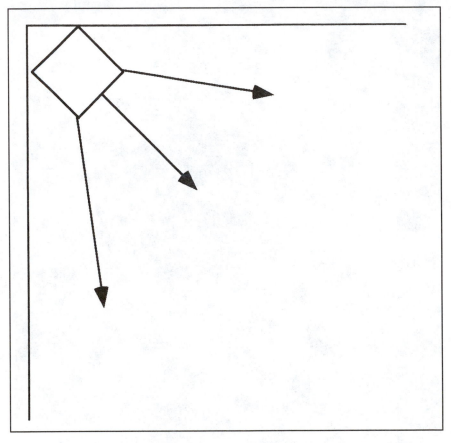

**Figure 9–3.** An example of correct loudspeaker positioning. Note the potential for increasing reflected sound energy.

An empty classroom might sound very different from an occupied classroom. That is, we typically install a sound-field FM unit in an empty room. When students arrive with their sound absorbent bodies and their noise-generating abilities, the sound-field FM unit might need to be readjusted.

## DURABILITY, FLEXIBILITY, PORTABILITY, AND EASE OF INSTALLATION OF SOUND-FIELD FM EQUIPMENT

Some sound-field FM units are intended to be permanently installed on the walls or in the ceiling of a particular room, while

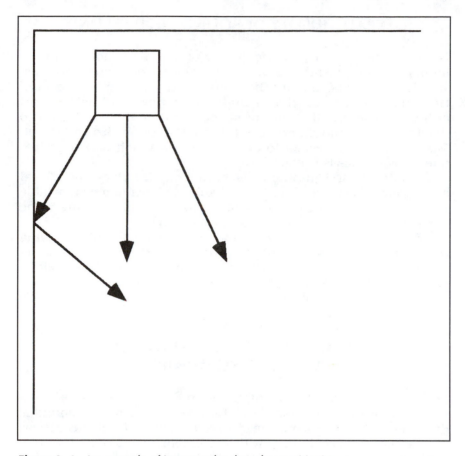

**Figure 9–4.** An example of incorrect loudspeaker positioning.

others need to be moved to other rooms or within a room as seating arrangements and learning centers change. Questions that need to be asked include: how user-friendly is the equipment? how heavy are the loudspeakers? how flexible is the speaker wire? how many hands do you need to transport the equipment? how robust is the sound-field FM equipment? (equipment that is fickle or fragile is of no value to anyone); does the equipment break easily? how quickly can repairs be made? are loaner parts available? is there a 30-day trial period? (Radio Shack does not have a trial period); and, overall, how supportive is the manufacturer?

## OVERALL FIDELITY OF SOUND-FIELD FM UNIT

Issues to consider include the quality of the microphone, and the loudspeaker frequency response and dispersion characteristics. We also need to be mindful that room acoustics shape the acoustic signal. It appears that a high-frequency emphasis signal enhances the intelligibility of speech and enables children to detect word/sound differences (Flexer, Millin, & Brown, 1990). More research is needed to determine how much of a high-frequency emphasis is needed.

Listen to the equipment. Does it crackle? Does the signal sound like it is on the verge of feedback when set to a comfortable listening level? Are there "hot spots" where feedback occurs if the teacher walks near certain areas in the room?

A sound-field FM system is meant to be a valuable teaching tool that facilitates classroom instruction thereby enhancing learning. If equipment malfunctions in any way or sounds "weird," or interferes with teaching, teachers likely will turn off the unit rather than fix the problem.

## PLACEMENT OF A CHILD WITH A KNOWN HEARING IMPAIRMENT

Technically, sound-field FM amplification amplifies the entire classroom to a relatively constant level. However, we know that sound is degraded as it is propagated away from the source (Leavitt & Flexer, 1991). Therefore, continuing with the pragmatic orientation of this book, it seems logical to seat the child who has a known hearing problem (typically minimal, mild, unilateral, or fluctuating) close to one of the loudspeakers (for sound enhancement), and close to the teacher/media materials (for visual enhancement).

## WHAT TYPE OF MICROPHONE SHOULD THE TEACHER WEAR?

Microphones used by FM systems are meant to be placed within six inches of the sound source. If they are placed further away, or if they are placed off axis, or if they are placed on noisy, movable surfaces such as on necklaces, sound distortion results. Figures 9–5 and 9–6 show examples of correct and incorrect microphone placement.

**Figure 9–5.** Correct microphone placement on the teacher (Courtesy of Audio Enhancement).

**Figure 9–6.** Incorrect microphone placement on the teacher (Courtesy of Audio Enhancement).

Typical microphone styles include: lavalier, collar, or boom (head-worn) types. The boom or head-worn microphone seems to offer the most consistent signal because the microphone turns when the teachers head turns. The disadvantage of this set-up is that teachers report that it can be uncomfortable. See Figures 9–7, 9–8, and 9–9 for examples of microphone styles.

## NECESSITY OF A "POINT-PERSON"

A critical variable for the effective installation and use of sound-field FM equipment is the presence of a support/contact person within or easily available to the school district who can install the equipment, present inservice training, and monitor equipment function and use. Classroom amplification systems represent a significant change for school systems both in operation and philosophy; that change requires a facilitator. Without a facilitator/support person, sound-field FM equipment might not be used or might not be used efficiently.

**Figure 9–7.** Lavalier microphone (Courtesy of Audio Enhancement).

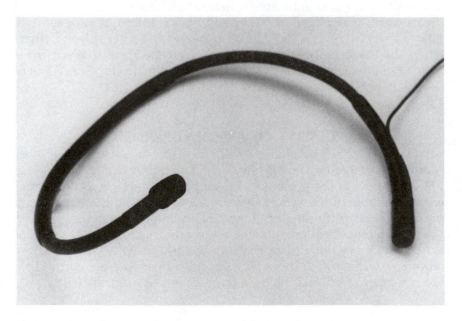

**Figure 9–8.** Collar microphone (Courtesy of Audio Enhancement).

## ACOUSTIC FEEDBACK CONTROL

Acoustic feedback in a sound-field system occurs when the amplified signal from a loudspeaker is picked up via the wireless microphone and is then "re-amplified." This re-amplification of sound creates system oscillation, which perceptively results in a "howl" or "squeal" to the listener. Acoustic feedback with sound-field amplification can occur when: (1) the teacher is positioned too close to a loudspeaker; (2) the gain, or volume, of the system is set too high; (3) or both. Thus, to avoid feedback, loudspeakers must be positioned so that the teacher can be within any instructional area in the classroom, but not in close proximity to the speaker. Loudspeakers positioned in the ceiling, elevated on the walls, or in noninstructional areas can fulfill this objective. In addition, highly directional loudspeakers can assist in acoustic feedback reduction.

To reduce the potential of acoustic feedback in the classroom, it is also imperative that the gain of the sound-field system not be set beyond levels that would amplify the teacher's voice by more than 8–10 dB. Gain settings that provide levels in excess of this

**Figure 9–9.** Head-worn Boom microphone (Courtesy of Audio Enhancement).

can result not only in feedback, but also in reductions in signal quality. The gain of a sound-field system also can be increased (i.e., reduce the potential of acoustic feedback) by decreasing the distance from the teachers mouth to the microphone (a boom microphone is particularly useful in reducing this distance) or by using a highly directional microphone. A unidirectional wired microphone could also be used to reduce acoustic feedback, however, such a set-up will restrict teacher mobility (Berg, 1993).

Operationally, the maximum gain of a sound-field unit (before reductions in quality and/or feedback occur) can be determined via the following formula (Berg, 1993):

$$\textbf{Maximum Gain (in dB)} = \textbf{20 Log } D_0 - \textbf{20 Log } D_s$$
$$+ \textbf{ 20 Log } D_1 - \textbf{20 Log } D_2 - \textbf{6 dB}$$

where $D_0$ = distance from the speaker to the listener; $D_s$ = distance from speaker to the microphone; $D_1$ = distance from the loudspeaker to the microphone; and $D_2$ = distance from the loudspeaker to the listener. Thus, if we assume that $D_0$ (distance from the speaker to the listener) is 20 feet; $D_s$ (distance from speaker to the microphone) is 0.5 feet; $D_1$ (distance from the loudspeaker to the microphone) is 5 feet; and $D_2$ (distance from the loudspeaker to the listener) is 10 feet, we develop the following equation for maximum gain:

$$\textbf{20 dB Maximum Gain} = \textbf{20 Log 20} - \textbf{20 Log 0.5}$$
$$+ \textbf{20 Log 5} - \textbf{20 Log 10} - \textbf{6 dB}$$

## SUMMARY

This chapter has presented many issues to consider when amplifying a classroom. A pragmatic approach is recommended. The goal of sound-field FM amplification is to provide an even and consistent S/N ratio improvement of approximately 10 dB throughout the learning area. The learning area to be amplified depends on the physical layout of the classroom, the listening demands placed on the pupils, and the teacher's individual instructional style. The reader is referred to Appendix B for use and maintenance tips for sound-field FM amplification systems and to Appendix C for troubleshooting tips.

# REFERENCES

Berg, F. (1993). *Acoustics and sound systems in schools.* San Diego, CA: Singular Publishing.

Boothroyd, A. (1992). The FM wireless link: An invisible microphone cable. In M. Ross, (Ed.), *FM auditory training systems: Characteristics, selection and use.* Timonium, MD: York Press.

Flexer, C., Millin, J.P., & Brown, L. (1990). Children with developmental disabilities: The effect of soundfield amplification on word identification. *Language, Speech and Hearing Services in Schools, 21,* 177–182.

Leavitt, R., & Flexer, C. (1991). Speech degradation as measured by the rapid speech transmission index (RASTI). *Ear and Hearing, 12,* 115–118.

# APPENDIX A
# SOUND-FIELD FM AMPLIFICATION MANUFACTURERS

**Anchor Audio, Inc.**
913 West 223rd Street
Torrence, CA 90502
(310) 533-5984

**Audio Enhancement**
1748 West 12600 South
Riverton, UT 84065
(801) 254-9263

**Custom Audio Design**
Box 597
Wenatchee, WA 98807-0598
(800) 355-7525

**Lifeline Amplification Systems**
55 South 4th Street
Platteville, WI 53818
(800) 236-4327

**Phonic Ear, Inc.**
3880 Cypress Drive
Petaluma, CA 94954-7600
(707) 769-1110

**Telex Communications, Inc.**
Educational Products
9600 Aldrich Avenue South
Minneapolis, MN 55420
(800) 328-3102

# APPENDIX B
# USE AND MAINTENANCE TIPS FOR
# SOUND-FIELD SYSTEMS

- Handle the equipment carefully to avoid bumping, jarring, or dropping.
- Avoid placing the FM equipment in areas of excessive cold or heat such as in direct sunlight.
- Avoid placing equipment in areas where it might be damaged by water, excessive dust, or moisture (e.g., near an open window, sink or aquarium; near the chalkboard).
- Each morning remember to turn on the receiver/amplifier and transmitter/microphone and check the operation of the FM system. If there is no sound, decreased volume or poor sound quality, use the troubleshooting checklist to identify the problem.
- Do not cut, pin or staple through the microphone connector cord or the speaker wires.
- Remember to turn off all equipment (transmitter/microphone and receiver/amplifier) at the end of each day.
- Remember to charge the transmitter each night. (Do not charge disposable batteries as they may rupture and damage the unit.)
- Remember to insert batteries with the correct polarity (+ to + and – to –).
- Position the transmitter face down on a cushion or other soft surface when removing and inserting batteries.
- Provide a dust cover to protect the microphone/transmitter and receiver/amplifier at night and over weekends and holidays.
- Periodically clean the equipment with a soft cloth. Do not spray any cleaning agents on the equipment.

---

Florida Department of Education, 1994. Reprinted with permission.

# APPENDIX C
# TROUBLESHOOTING SOUND-FIELD SYSTEMS

## Problem: No FM Reception (The NO FM light stays on.)

1. Verify that the frequency code number of the transmitter and the receiver/amplifier are the same.
2. Verify that the frequency selector switch is set correctly.
3. Check battery's charge and polarity (+ 10 + and – 10 –). Recharge or replace the transmitter batteries.
4. Verify that the antenna is properly connected to the antenna jack on the receiver/amplifier.
5. Contact the audiologist if you have checked all of the above and the NO FM light remains illuminated.

## Problem: No Sound

1. Check items 1–4 under NO FM Reception.
2. Verify that the receiver/amplifier is turned to the ON position.
3. Verify that the transmitter is turned on.
4. Check connections between receiver/amplifier and wall transformer.
5. Plug the receiver/amplifier into a different wall outlet.
6. Verify that speaker wires are connected to the receiver/amplifier and speaker terminals.
7. Verify that speaker wire leads are not touching each other at the speaker or the receiver/amplifier terminals.
8. Check the microphone connector cord for damage. (Wiggle the cord to determine if wires have been damaged.)
9. Contact the audiologist if you have checked all of the above and there is still no sound from the system.

## Problem: Weak Speaker Output

1. Recheck wiring to make sure cables are connected in phase.
2. Recharge and/or replace battery.
3. Check the receiver/amplifier volume control setting(s).
4. Reposition microphone closer to the speaker's mouth.

---

Florida Department of Education, 1994. Reprinted with permission.

5. Realign speakers if they have been moved so they are no longer directed toward the students.
6. Contact the audiologist if you have checked all of the above and system continues to have weak speaker output.

## Problem: Poor Sound Quality

1. Replace battery.
2. Check battery prong connections in the transmitter and charger unit. (The + and − connections in the battery compartment and the charger unit need to be periodically tightened to ensure proper contact.)
3. Verify that the antenna is properly attached to the receiver/amplifier.
4. Check the speaker wires for cuts or poor connections.
5. Check the microphone connector cord for damage. (Wiggle the cord to determine if wires have been damaged.)
6. Check the volume control settings on the receiver/amplifier.
7. Contact the audiologists if you have checked all of the above and the system continues to exhibit poor sound quality.

## Problem: Feedback

1. Lower the FM volume control setting on the receiver/amplifier.
2. Increase the distance between the speaker(s) and the transmitter microphone.
3. Contact the audiologist if you have checked the above items and the rophone system continues to feed back.

## Problem: FM Interference

1. Contact the audiologist and describe the type of interference noted (e.g., paging system, security system, other classroom conversation, etc.)
2. A different FM frequency will need to be selected for the classroom in the event of interference from another broadcast source.
3. The receiver/amplifier will need to be reset to accommodate the change in transmitter frequency.

---

Florida Department of Education, 1994. Reprinted with permission.

# CHAPTER
# 10

# INSERVICE TRAINING FOR THE CLASSROOM TEACHER

*Gail Rosenberg*
*Patricia Blake-Rahter*

Classroom teachers who have used sound-field FM amplification systems have enjoyed a new relationship with their students. These systems are easy to use, eliminate the acoustic barriers of noise, reverberation, and distance, and help to improve student listening behaviors. **In order to promote this unique technology effectively, it is essential to conduct inservice training for participating classroom teachers prior to implementing a sound-field FM amplification program.** Not only is it important to orient teachers to equipment use, but it is also necessary to provide information about the listening process, classroom acoustics, the effects of noise and reverberation on speech perception, classroom modifications, and listening strategies. Successful use of a sound-field FM amplification system in the classroom is contingent on the quality of the inservice training and technical support available when the equipment is in use. This chapter provides an overview of components of an inservice training module for classroom teachers that was developed through the Improving Classroom Acoustics (ICA) project (Florida Department of Education, 1994). Additional references are provided for each major section to assist inservice trainers in obtaining collateral information. Table 10–1 shows an outline for providing inservice training for classroom teachers and other interested professionals who will be using sound-field FM amplification.

**Table 10–1.** Outline for teacher inservice training on sound-field FM amplification.

---

**CONTENT AREAS**

---

**I.  The Importance of Listening**
   A.  Hearing vs Listening
     1.  Definition and Differentiation of Terms
     2.  Sequential Levels Involved in the Listening Process
   B.  Variables that Affect Listening
     1.  Acoustic Signal Variables
     2.  Listener Response Task Variables
     3.  Listening Environment Variables
     4.  Listener-related Variables
     5.  Speaker-related Variables

**II.  Orientation to Sound and Its Properties**
   A.  Sound and Sound Waves
   B.  Characteristics of Sound
     1.  Frequency
     2.  Intensity
     3.  Time
   C.  Critical Frequencies for Speech Perception
   D.  Low and High Frequency Characteristics of Speech Sounds
   E.  Activity: Unfair Spelling Test

**III.  Factors that Limit the Student's Ability to Listen**
   A.  Noise
   B.  Reverberation
     1.  Reverberation and Reverberation Time
     2.  Effect of Reverberation on Speech Perception
     3.  Effect of Age on Speech Perception Under Reverberant Conditions
   C.  Effects of Noise and Reverberation on Speech Perception
     1.  Synergistic Phenomenon of Noise and Reverberation
     2.  Acceptable Acoustical Standards for Classroom Noise and Reverberation Levels
   D.  Signal-to-Noise (S/N) Ratio, Distance, and Speech Recognition
     1.  Signal-to-Noise Ratio (S/N)
     2.  Effect of S/N and Distance on Speech Recognition Ability as a Function of Listener Age

**IV.  Noise and Reverberation Sources**
   A.  External and Adjacent Classroom Noise Sources
   B.  Internal Classroom Noise Sources
   C.  Common Classroom Noise Levels
   D.  Common Environmental Noise Levels
   E.  Reverberation Sources

**V.  Materials and Strategies to Manage and Enhance Listening**
   A.  Noise and reverberation absorption materials
   B.  Acoustical treatments
     1.  External Modifications

---

2. Internal Modifications
   C. Instructional strategies to improve classroom listening
   D. School-based noise reduction strategies

**VI. *Sound-field FM Amplification***
   A. Orientation to Sound-field FM Amplification
      1. Description of Components
      2. Demonstration of Sound-field FM Amplification System
      3. Benefits of Sound-field FM Amplification
      4. Factors to Consider When Recommending, Purchasing, Installing, and Using Sound-field FM Amplification Systems
   B. Transmitter and Microphone Options
      1. Options for Wearing the Transmitter/Microphone
      2. Tips for Successful Use of the Transmitter/Microphone
   C. Speaker Options
      1. Speaker Placement Options
      2. Speaker Placement Variables
   D. Setting the Receiver/Amplifier Volume Control
   E. Sound-field FM Amplification System Maintenance Tips
   F. Sound-field FM Amplification System Troubleshooting Tips (wired system)

**INSERVICE TRAINING MATERIALS**
- Overhead Projector
- Cassette Tape Player
- Unfair Hearing Test Cassette Tape (Collins, 1989)
- Sound-Level Meter
- Sound-field FM Amplification System (installed) with available microphone options
- Auxiliary input cords and adapters
- Participant notebooks (copies of transparencies and pertinent reference items such as troubleshooting and maintenance checklists, equipment and support literature from the manufacturer, etc.)

# THE IMPORTANCE OF LISTENING

Listening is a dynamic and complex perceptual process. Listening is very simply defined as "hearing others talk and attending to and trying to understand what they are saying" (Berg, 1993, p. 1). As such it is a major component of the communication network. Butler (1975) estimated that listening comprises 45 percent of the typical daily communication for adults. However, young children may actually spend as much as 60 percent of their school day involved in the listening process. When children experience listening difficulty because of fluctuant or permanent hearing loss or if there are any barriers present such as noise, reverberation, or distance from the teacher, students' opportunities for efficient and effective learning may be compromised. Education in the 1990s has demonstrated that students are expected to retain more complex information at an earlier age (Florida Commission on Education Reform and Accountability, 1992). Students who experience difficulty at any level in the listening process will find it more problematic to use auditory information in an efficient and effective manner. The reader is referred to Chapter 11 for additional information on the importance of listening.

Classroom teachers may find that student and/or observation checklists provide vital information about students who are experiencing classroom listening difficulties. A listing of available observation forms is provided in Appendix A. Assistance from the school's speech-language pathologist or audiologist may be requested in obtaining and completing these forms. Students with persistent listening problems should be referred to the school's student study team for assistance.

Additional references on the listening process include: Anderson (1989b), Berg (1987), DeConde (1984), Downs and Crum (1978), Edwards (1991a, 1991b, 1993), Flexer (1992a, 1993), Finitzo (1988), Florida Department of Education (1988, 1991c, 1991d, 1994), Keith (1988), Kent (1992), Lasky (1983), Lasky and Chapandy (1976), Lasky and Cox (1983), Lasky and Tobin (1973), Neuman and Hochberg (1983), Papso and Blood (1989), Rabbit (1966), Rupp (1986), Rupp, Jackson, and McGill (1986), Sanders (1985), and Smoski, Brunt, and Tannahill (1992).

## ORIENTATION TO SOUND AND ITS PROPERTIES

A prerequisite for understanding the elements that deleteriously affect our ability to hear and listen is to achieve a working knowledge of sound and its properties. Sound is produced by the movement or vibration of molecules in the air, and the sounds we hear are essentially vibrational energy. When a person speaks or when sound is produced in some other way, displacement or disruption of air particles occurs and this motion creates sound waves. Sound waves emanate from the sound source to many different locations in multiple directions, traveling through the air at a speed of approximately one-fifth of a mile per second (1,130 feet per second). The three physical characteristics of sound are frequency (pitch), intensity (loudness), and time (duration). Frequency is very simply described as the number of oscillations or cycles produced by a sound source within a given time while the sound is produced. Cycles per second (cps) or Hertz (Hz) are reference terms for frequency. For instance, if a vibrator (sound source) were set into motion and completed 1,000 back-and-forth cycles in one second, it would have a frequency of 1000 Hz. Time is usually measured in seconds or milliseconds (msecs). The Familiar Sounds Audiogram (Northern & Downs, 1991) illustrates that the lower frequencies represent fewer cycles per second (see Figure 10–1). As the frequency of sound increases (e.g., from 250 to 8000 Hz), wavelengths become shorter and the number

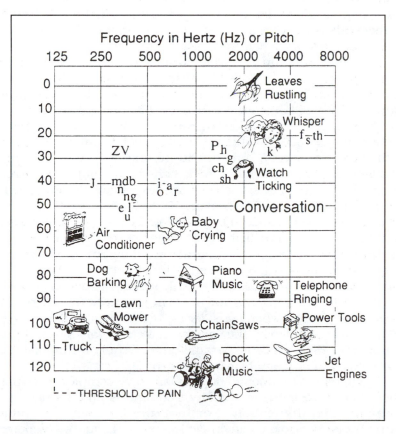

**Figure 10–1.** The Familiar Sounds Audiogram shows the approximate frequency and intensity levels for common environmental sounds. Figure used with permission from Northern, J., & Downs, M. (1991). *Hearing in children* (4th ed.). Baltimore: Williams & Wilkins Co.

of cycles per second increases. Correspondingly, a sound's wavelength always increases as frequency decreases.

The human ear can perceive a wide range of frequencies and is especially sensitive and responsive to vibrational energy between 250 and 4000 Hz. However, vibration rates between 16 and 23,000 Hz may be heard by some persons (Berg, 1986). Frequencies between 300 and 3000 Hz are particularly critical for the perception of speech (Roeser, 1988).

The second characteristic of sound is intensity or the loudness of sound. A sound's intensity is determined by the degree of movement or displacement of air particles that occurs as a sound is created. Intensity represents the strength of sound waves emanating from the sound source and is expressed in decibels (dB). A sound that is 0 dB is at a loudness level that a normal listener can just barely hear. The decibel scale is complex and logarithmic so that each increase of 10 dB indicates a tenfold increase in sound intensity.

Just like common everyday sounds we hear in our environment, speech sounds each have frequency, intensity, and time characteristics. Figure 10–2 shows that most of the energy of vowel sounds is located in the low frequencies (60–500 Hz) and the energy of consonant sounds is concentrated in the mid and higher frequency range (1000–8000 Hz). Vowel sounds comprise 60 percent of the power of speech but contribute only 5 percent toward the intelligibility of speech. Conversely, consonant sounds represent 5 percent of the power of speech but are responsible for 60 percent of the intelligibility of speech (Leavitt, 1978; Madell, 1990). Therefore, the greatest power of speech is in the low frequencies and the greatest information for intelligibility is concentrated in the higher frequencies. All sounds are also characterized by time (i.e., they are of long or short duration). Speech sounds are relatively short in duration (usually less that 300 msec), while some environmental sounds can be quite long lasting for several minutes or hours. Time becomes important when considering the short duration of speech sounds co-articulated by the teacher in a classroom situated near a heater vent. Here, all three parameters of frequency, intensity, and time come into play and heavily affect the S/N ratio.

Common environmental sounds are very complex in structure since they are comprised of many frequencies and intensity levels occurring simultaneously. For instance, speech is a complex sound. Noise is also a common environmental sound. The power of speech and noise represent a reverse relationship. Noise levels are typically greater in the low frequencies and then become progressively less powerful at the higher frequencies. As discussed in

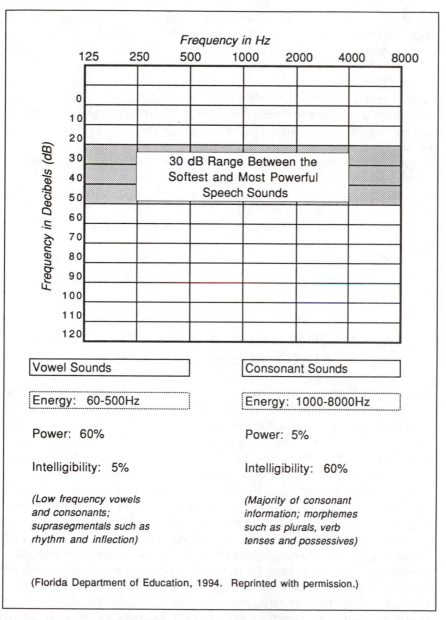

**Figure 10-2.** Low and high frequency characteristics of speech vowel and consonant sounds are shown with their contributions to the power and intelligibility of speech. Figure used with permission from Florida Department of Education (1994). *Improving classroom acoustics: Inservice training manual.* Tallahassee, FL.

Chapter 3, noise affects the intelligibility of consonant sounds more markedly than it does vowel sounds. This occurs because the more powerful noise in the low frequencies tends to mask the less powerful high-frequency speech sounds that are critical for speech recognition. Noise indeed acts as an acoustic filter in this instance. High-frequency sounds such as /s/, /z/, or /t/ are masked or distorted because they are relatively weak, short in duration, and ultimately more highly susceptible to the effects of noise than are most vowel sounds. Consonant sounds are also distorted by noise because they include more high-frequency energy which is absorbed by most room surfaces. The authors have found that administration of the recorded unfair spelling test (Collins, 1989) under mild hearing loss conditions, such as with hearing protection, is useful in demonstrating to participants a simulated distorted speech listening experience.

Additional references on orientation to sound and its properties: Cherow (1986), Collins (1989), Florida Department of Education (1988, 1991a, 1994).

## FACTORS THAT LIMIT THE STUDENT'S LISTENING AND PERCEPTUAL ABILITIES

Noise, reverberation, distance, and their synergistic interactions are acoustic barriers that serve to limit a student's ability to listen effectively in the school environment (see Chapter 3). A classroom should be sufficiently quiet to permit a teacher's voice to be heard with little difficulty (Ross, 1972). Classroom noise is a variable that most assuredly influences listener effectiveness and often produces teacher fatigue. Similarly, young students attempting to learn new information very likely find that classroom noise renders learning a stressful experience.

The combined effect of noise and reverberation on speech perception is of paramount concern in the classroom listening environment. While it is easy to note the presence of noise, the presence of even moderate reverberation is not readily apparent to most listeners with normal hearing. As a result, the subtleties of reverberation can produce frustrated listeners who exhibit an inability to understand speech clearly (Nabelek & Robinson, 1982). Listening difficulties escalate when noise and reverberation combine in a synergistic manner (Crandell & Bess, 1986; Finitzo-Hieber & Tillman, 1978; Nabelek & Pickett, 1974). Research studies on the cumulative effects of noise, reverberation, and distance have shown that students seated in the middle to rear portion of a typical classroom have greater difficulty understanding speech than traditionally has been suspected (Leavitt & Flexer, 1991). For maximum communication to occur, acceptable acoustical standards must be met. Unoccupied classroom noise levels should not exceed 35 dB(A) and reverberation levels should not surpass 0.4 second (Crandell, 1991b).

Suggested activities to accompany this discussion include: (1) demonstration of a sound-level meter which is helpful in illustrating signal-to-noise (S/N) ratio; and (2) participants' identification of noise sources in their personal and classroom environments (see suggested strategies offered later in this chapter). Refer to Chapters 3, 4, and 5 for additional information on noise and reverberation studies related to classroom acoustics and the effects these acoustic barriers impose on speech perception ability.

Additional references on the effects of noise, reverberation, and signal-to-noise on speech perception: Anderson (1989b), Berg (1986, 1987, 1991), Bess and Logan (1984), Bess and McConnell (1981), Blair (1977), Crandell (1992, 1993b, 1993c), Crandell and Bess (1987b), Dirks, Morgan, and Dubno (1982), Edwards (1993), Elliot (1979, 1982), Fior (1972), Finitzo (1988), Flexer (1992a), Florida Department of Education (1988, 1991b, 1991d, 1994), Gengel (1971), Madell (1990), Markides (1986), McCroskey and Devens (1975), Neuman and Hochberg (1983), Neuss, Blair, and Viehweg (1991), Olsen (1981), Phonic Ear, Inc., (1993a), Richards, Flexer, Brandy, and Wray (1994), Ross (1978, 1992), Ross and Giolas (1971), Sanders (1985), and Yacullo and Hawkins (1984).

## NOISE AND REVERBERATION SOURCES

As discussed in Chapter 3, a noise source may be external (outside the building or classroom) or internal (within the classroom itself). Classroom noises are generated by four basic classroom functions: (1) movement (of children, teachers, or furniture on hard surface floors); (2) equipment (noisy or malfunctioning equipment such as HVAC systems, fans, audiovisual equipment, clock, computer terminal, fluorescent lighting, etc.); (3) activities (teaching and learning activities occurring simultaneously in the same classroom); and (4) facilities (noise from a bathroom, computer terminal, or other special purpose area in the classroom).

In addition to these sources of noise there are other elements within the classroom itself that are contributors to the acoustic climate and they include: age and number of students in the classroom; classroom configuration; height and type of ceiling; type of floor surface(s); interior wall surfaces; interior doors; windows; furniture arrangement; and classroom noise, both incidental and that generated during learning activities. The ICA inservice training manual (Florida Department of Education, 1994) contains a Classroom Description Worksheet that provides the teacher a comprehensive tool for noting all possible noise sources as well as acoustical treatments (see Appendix B). Berg (1993) also provides a number of checklists for identifying noise sources and acoustical modifications.

Research studies show that the recommended acoustical level of 35 dB(A) for unoccupied classrooms is rarely achieved. Berg (1993) describes a range of typical classroom noise levels: 30–35 dBA, unoccupied at night; 40–50 dBA, unoccupied with the heating/ventilation/air conditioning (HVAC) system in operation; and 55–75 dBA, occupied by 25 students and a teacher. However, he stresses that for students to listen effectively the unoccupied noise level should not exceed 35–40 dBA and 40–50 dBA for an occupied classroom. It is also important to note that over the past several decades, ambient noise levels in traditional unoccupied classrooms have shown relatively little change. Classroom and school environment noise studies are producing an ever-growing body of evidence that emphasizes the need for improvement in the acoustical environments in which learning occurs. Reverberation in classrooms results from noise and acoustic signals coming in contact with hard, nonporous, and nonelastic surfaces. Flat surfaces such as smooth, unpainted concrete walls and ceilings, bare floors, and large expanses of regular pane windows are highly reverberant.

In addition to noise in the school environment, most children are exposed to noisy environments at home. Downs and Crum (1978) reported that noise in home or educational environments may negatively affect the learning performance of children. Noise levels of common environmental sounds and their effects on humans are given in Table 10–2. (Refer to Chapter 3 for additional information on noise and reverberation sources in the educational environment.)

Additional references on noise sources, classroom noise levels, and reverberation sources: Berg (1986, 1987, 1991), Borrild (1978), Crandell (1991b), Finitzo (1988), Florida Department of Education (1988, 1991d, 1994), Hammond (1991), Rosenberg, Blake-Rahter, Allen, and Redmond (1994), and Ross and Giolas (1971).

**Table 10–2.** Intensity levels for common environmental sounds as well as their effects.

| Sound | Noise Level (dB) | Effect |
|---|---|---|
| Boom Cars | 145 | |
| Jet Engines (Near) | 140 | |
| Shotgun Firing | 130 | |
| Jet Takeoff (100–200 ft.) | 130 | |
| Rock Concerts (Varies) | 110–140 | Threshold of pain (125 dB) |
| Oxygen Torch | 121 | |
| Discotheque/Boom Box | 120 | Threshold of sensation (120 dB) |
| Thunderclap (Near) | 120 | |
| Stereos (Over 100 Watts) | 110–125 | |
| Symphony Orchestra | 110 | |
| Power Saw (Chain Saw) | 110 | Regular exposure of more than 1 min. risks permanent |
| Pneumatic Drill/Jackhammer | 110 | hearing loss (over 100 dB) |
| Snowmobile | 105 | |
| Jet Flyover (1000 ft.) | 103 | |
| Electric Furnace Area | 100 | |
| Garbage Truck/Cement Mixer | 100 | No more than 15 min. unprotected exposure recommended |
| Farm Tractor | 98 | (90–100 dB) |
| Newspaper Press | 97 | |
| Subway, Motorcycle (25 ft) | 88 | Very annoying |
| Lawnmower, Food Blender | 85–90 | Level at which hearing damage (8 hrs.) begins (85 dB) |
| Recreational Vehicles, TV | 70–90 | |
| Diesel Truck (40 mph, 50 ft.) | 84 | |
| Average City Traffic Noise | 80 | Annoying; interferes with conversation; constant exposure |
| Garbage Disposal | 80 | may cause damage |
| Washing Machine | 78 | |
| Dishwasher | 75 | |
| Vacuum Cleaner, Hair Dryer | 70 | Intrusive; interferes with telephone conversation |
| Normal Conversation | 50–65 | |
| Quiet Office | 50–60 | Comfortable (under 60 dB) |
| Refrigerator Humming | 40 | |
| Whisper | 30 | Very quiet |
| Broadcasting Studio | | |
| Rustling Leaves | 20 | Just audible |
| Normal Breathing | 10 | |
| | 0 | Threshold of normal hearing (1000–4000 Hz) |

*Source:* Sources include the American Medical Association and the Canadian Hearing Society of Ontario. Decibel table developed by the National Institute on Deafness and Other Communication Disorders, National Institutes of Health, Bethesda, Maryland 20892. January 1990.

Since the sensitivity of the ear to sound is not the same for all frequencies, weighting or attenuating filters are included in the sound level meter's circuits to simulate the ears' response. A noise level meter gives an instantaneous measurement of the noise present, but cannot measure the duration of the exposure. To measure the amount of noise a person is exposed to over a period of time, a "dosimeter" or an Integrated sound level meter must be used.

This decibel (dB) table compares some common sounds and shows how they rank in potential harm to hearing.

In many industries, workers are exposed to dangerous levels. This is particularly true in the construction, lumber, mining, steel and textile industries.

## MATERIALS AND STRATEGIES TO MANAGE AND ENHANCE LISTENING AND PERCEPTION

Managing sound and noise control in the classroom and school environment is both a science and an art. All students and the teacher are able to enjoy a more pleasant listening, learning, and teaching experience when measures have been taken to reduce noise and reverberation sources in the classroom. These efforts also yield improved morale and motivation by reducing stress generated by a noisy, reverberant learning environment.

Materials are available within school and classroom environments that can absorb sounds and assist in decreasing the effects of noise and reverberation on speech perception. The reader is directed to Chapter 6 for a discussion of acoustical modifications in the classroom.

Additional references on acoustical treatment: Berg (1986, 1987, 1993), Berry (1988), Bess and Logan (1984), Bess and McConnell (1981), Bess, Sinclair, and Riggs (1984), Bloomfield (1987), Crandell (1991a, 1991b), Crandell and Smaldino (1992), Crandell, Smaldino, and Flexer (1994), Crum and Matkin (1976), Finitzo (1988), Florida Department of Education (1988, 1991d, 1994), Hart (1983), Jones, Berg, and Viehweg (1989), Leggett, Brubaker, Cohodes, and Shapiro (1977), Olsen (1981), Ross (1978, 1992), and Schneider (1992).

## INSTRUCTIONAL STRATEGIES TO IMPROVE CLASSROOM LISTENING AND PERCEPTION

*Throughout all levels of our educational development, listening is the main channel of communication and thus plays a significant role in our educational, personal, and professional lives (Peters & Austin, 1985, p. 97).*

Ease of listening requires good acoustics, a comfortably loud speech signal, familiar language, easily understandable content, and an appropriate pace at which new information is presented (Sanders, 1985). A listening problem arises when any of these components exceeds an individual's tolerance or proficiency level. Implementation of instructional listening behavior strategies will benefit students and the teacher particularly in less than optimum acoustic environments.

Listening strategies involve temporary alteration by the speaker or the listener of a characteristic of the listening environment (Edwards, 1991a). For ideal speech communication to occur, the distance between the teacher and student(s) should be only 3–4 feet (Harris, 1979). In the early grades the teacher serves as manager of the listening environment and in that role initiates most of the important listening strategies. However, as students develop the metacognitive capacity necessary to identify an awareness of difficult listening situations, the student must be the initiator of desirable listening strategies. Some students require instruction on the behaviors necessary to become an effective and efficient listener. The following classroom listening enhancement tips should be of interest to classroom teachers in structuring the classroom to create a positive listening environment. A classroom environment must invite the development and application of appropriate speaker and listener strategies. The following strategies are suggested.

- *Develop a sensitivity for classroom noise control.* Student activities that will enhance their awareness of noise generated by themselves and their classroom activities will identify and reinforce their efforts to reduce classroom noise levels. Provide classroom activities to sensitize students to the devastating effects of internal classroom noise to achieve positive attitudinal change.
- *Focus on structured learning and effective planning.* Schedule time for activities such as sharpening pencils, computer use, and restroom breaks to encourage increased time on task and to minimize the random, spontaneous movements of students during instructional periods. This also allows every student to have a "listening break" at the same time.
- *Maintain reduced noise levels during academic teaching.* An instructional setting where the teacher's voice is approximately 30 dB above the background noise level will produce an optimum listening situation. To be able to hear all of the consonant sounds in the acoustic signal the teacher's voice needs to be 33 dB above the ambient noise level. To receive all vowel sounds a range of 22 dB is required (Ross, Maxon, & Brackett, 1982).
- *Schedule instruction in the high academic content subjects earlier in the school day.* This allows students to receive information when they are more alert and are able to pay closer attention. Students with any degree of hearing or listening problem often become irritable or distractible later in the day because of the stress involved in listening and attending for supplemental visual clues. This situation may

be applicable to most student listeners in noisy classroom environments.

- *Encourage students to assist with enforcing classroom rules.* These rules governing student conduct and class discussion help to reduce the classroom noise level, especially if students are stakeholders and have input in development of the rules.
- *Keep classroom doors and windows closed.* If doors to areas outside the classroom remain closed, the amount of adjacent and external noise which may travel to the classroom will be decreased.
- *Use a well modulated speaking voice.* The intensity level of the teacher's voice as well as the rate of speech and modulation used when presenting information are critical factors for successful listening and comprehension by students. Having to compete with background noise generally results in frustration, voice fatigue, and an increased classroom noise level. Speaking at a moderate conversational intensity should allow students to hear the teacher easily in a typical classroom setting.
- *Use listening cues to encourage good listening skills.* Using a phrase or action to cue students to listen is valuable when teaching students with listening problems or in a less than optimal acoustic setting. Examples of these techniques include switching the lights off and on, using repetitive phrases such as "Are you ready?" or "It's time to listen." or using a particular action designated to gain students' attention. Encourage students to identify listening strategies that are helpful to them.
- *Move closer when speaking to students who appear to be experiencing listening difficulty.* A distance of 3–4 feet is optimum for listening, but moving even closer may be necessary for some students. This is particularly important for students presenting with fluctuant hearing loss caused by otitis media. In addition to moving closer to the student, speaking to the student at his/her ear level will provide a clearer speech signal. This maneuver will be helpful in a competing noise listening situation.
- *Simplify verbal directions as much as possible and provide directions in both verbal and written form.* Long, involved verbal directions with multiple step commands become even more confusing in the presence of classroom noise. When

students only partially complete assignments or homework, it may be that students have not clearly heard or understood the verbal directions. Some students simply are unaware that they have missed some information.

- *Use audio-visual teaching aids.* Use this strategy to supplement verbal instruction when providing important topical information and key vocabulary and concepts.
- *Use clarification strategies.* Effective strategies to enhance listener understanding include restating or rephrasing information, using listening cues previously described, and allowing students adequate time to integrate information prior to responding.
- *Monitor student comprehension and performance.* Welcome feedback by regularly using effective questioning and observation techniques. Use activities designed to teach students how to communicate acoustically clear and pragmatically appropriate messages, and reinforce students for using appropriate listening and speaking strategies. This will send a clear signal to all students that good communication is important.
- *Provide preferential and flexible seating.* Flexible preferential seating is essential for those students who appear to have difficulty listening (Edwards, 1993). Seating near the primary sound source and away from noisy areas or activities will provide the best seating and listening arrangement. Flexibility is important. As the position of the primary sound source changes, students requiring preferential seating should be allowed to move to another seat that affords the preferential seating advantage. Moving closer to targeted students when giving important information or instructions is helpful. It is interesting to note that a study conducted to measure the integrity of speech as it is propagated across the physical environment showed that the speech-like signal was substantially degraded even at the front-row center seat (Leavitt & Flexer, 1991). The signal received in the back-row center seat (row 7) showed a marked decrease of 45 percent of equivalent speech intelligibility.
- *Enable students to be good listeners.* Invite students to ask questions and request repetitions or clarification if they did not clearly understand information or instructions. Encourage students to be a "buddy" for another student who requires clarification strategies or additional listening clues.

Challenge students to help improve their listening environment by enforcing classroom rules.

- *Practice active, "whole body listening."* Emphasize the attending behaviors necessary to listen (e.g., being still, quiet, thinking about listening, and paying attention to sounds). Stress that good listening requires students to listen with their brain, ears, eyes, mouth, hands, and feet. When students concentrate on listening with various body parts it generally produces a quieting effect. Thus, they are listening with their whole body (Truesdale, 1990).

- *Reduce the noise level in the classroom at various times throughout the day.* This provides students a listening respite.

- *Invite the district's audiologist to make a presentation to the class.* The presentation should emphasize the importance of structuring and maintaining a good listening environment. Students could follow up with a special project on this topic.

- *Ask for assistance.* Consult with the audiologist, the school's speech-language pathologist, or with a student study team when a student is having considerable listening difficulty. It is important to rule out any sensory or other problems that may be contributing to the student's poor classroom listening behavior.

- *Use sound-field FM amplification in the classroom.* This system is a flexible audio enhancement tool that provides benefit to both the teacher and the students and it demands very little of the teacher using the system (Allen, 1993).

Additional references on strategies to improve listening skills: Berg (1993), Bess and Logan (1984), Bloomfield (1987), Conway (1990), Crandell (1991b), Edwards (1991b), Florida Department of Education (1988, 1991d, 1994), Finitzo (1988), Hart (1983), Ross, Maxon and Brackett (1982), Schneider (1992), and Silverstone (1982).

## SCHOOL-BASED NOISE REDUCTION STRATEGIES

School administrators and leadership teams may decide to take charge of noise and manage acoustical problems within the school environment. For example, a school-wide policy for noise control could be adopted. However, for this movement to be successful, students, staff, and parents must become involved in the noise abatement project. Support solicited from the Parent Teacher

Organization (PTO) could fund special projects such as landscaping, carpeting exterior hallways, or installing noise reduction strategies designed to promote the school's theme. Capital outlay funds may used to purchase necessary items to implement the approved plan. Additional avenues that school leadership teams may pursue to promote noise control include careful scheduling of classes and routing student traffic flow to decrease hallway noise and to ensure an ongoing preventative maintenance program. School leadership teams also have the opportunity to control noise through careful acoustical planning when a new school is to be constructed or when a new wing is added or an older site renovated. Acoustical needs should be matched with commercially available products offering Noise Reduction Coefficient (NRC) ratings and Sound Transmission Loss (STL) ratings to meet desired acoustical requirements (see Chapter 6). Also available are acoustical materials with specific acoustical diffraction qualities. The audiologist could be available to consult with the school and district level staff on acoustical concerns and new construction as a part of the planning team and also may be available to conduct noise measurements.

## SOUND-FIELD FM AMPLIFICATION SYSTEMS

### Orientation to the Sound-Field FM Amplification System

The basic components of a sound-field FM amplification system are illustrated in Figure 10–3. The stimulus is provided by the teacher via the microphone that is connected to the wireless transmitter. When sound enters the transmitter, the acoustic energy is converted to electrical energy and the transmitter sends the converted acoustic signal via an FM carrier wave to the amplifier. The transmitter is of a specific FM frequency and the amplifier is set to receive that FM frequency signal (components of the specific sound-field FM amplification system should be identified to the teacher at this time). The acoustic signal is then amplified approximately 8–10 dB and is broadcast to the students via portable loudspeakers that have been strategically placed in the classroom. Amplifying the teacher's voice more than 10–15 dB could introduce distortion into the FM system and would likely become a distraction to the students.

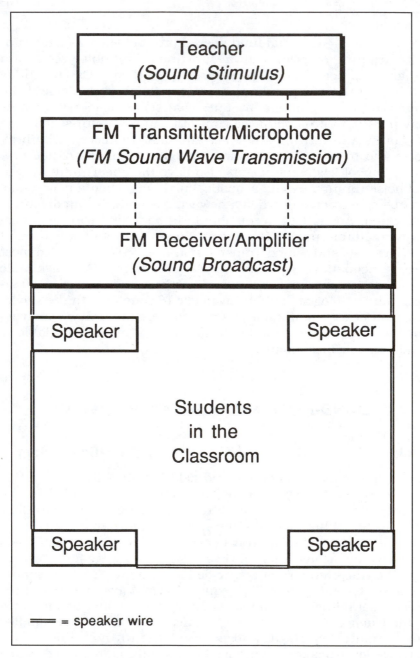

**Figure 10–3.** Block diagram of a sound-field FM amplification system shows components and the basic design of a four-speaker system. Figure used with permission from Florida Department of Education (1994). *Improving classroom acoustics: Inservice training manual.* Tallahassee, FL.

A sound-field FM amplification system offers a transmission range of up to approximately 300 feet. Since different FM frequencies are used for each classroom, there is little danger of picking up neighboring classroom conversation. There are 40 narrowband frequencies available for educational use (Federal Communications Commission, 1992). Other sound-field FM systems are available which employ VHF wideband frequencies (see Chapter 9). The constant level of the teacher's voice is controlled by setting the amplifier at a comfortable listening level, thereby allowing all students to be within the critical distance for optimum listening. With this listening arrangement all students are in effect closer to the teacher. Since students are closer to the sound source (loudspeaker), the acoustic signal suffers less degradation as it travels through the room. It has been shown that sound-field FM systems may be effectively used in self-contained, open plan, and portable classroom settings if correctly placed (Allen, 1993).

**A sound-field FM amplification system bridges the gap between the sound source and the listener. It not only enhances speech perception and reduces listening effort but it serves to increase the likelihood that students have the opportunity to maximize language and academic learning** (Ross, 1992). Primary advantages to using the sound-field FM amplification system are that it: projects the teacher's voice to a level at which students can hear comfortably without straining, improves the S/N ratio by producing a nearly uniform loudness level in the classroom which is unaffected by the teacher's position, and reduces the effects of noise, reverberation, and distance from the teacher so that students in the back of the classroom can hear the teacher's voice as clearly and precisely as can students seated near the teacher. Students also enjoy using the transmitter/microphone for sharing time and giving reports to the class. Adapters are available so students with personal FM systems may use these assistive devices in conjunction with the sound-field FM amplification system. If the systems are compatible (same frequency) there is no need for an adapter.

Prudent consideration must be given to student, school, and classroom needs prior to recommending, purchasing, installing, and using a sound-field FM amplification system. Refer to the checklist in Appendix C for additional information about key factors to review when considering use of a sound-field system. These factors include: school location and classroom facilities, classroom usage and teaching style, frequency specifications, equipment needs and options, equipment manufacturer considerations, selection capabilities criteria, cost effectiveness, administrative support, funding options, and ongoing maintenance. The true benefits of S/N technology become apparent after the sound-field FM system has been effectively installed and functions properly with high fidelity, and when a support person is easily accessible, and if the school's leadership team recognizes the value of hearing (Flexer, 1992a).

Research has shown that sound-field FM amplification may be of benefit to a number of pediatric populations with presumably "normal" or minimally impaired hearing sensitivity who are at risk for speech recognition difficulties in degraded listening environments (Benafield, 1990; Crandell & Bess, 1987b; Crandell & Smaldino, 1992; Elliott, 1982; Flexer, 1992a, 1993; Flexer, Millin, & Brown, 1990; Nabelek & Nabelek, 1985, Neuss, Blair, & Viehweg, 1991). Among these at risk populations are students who present with learning disabilities, fluctuant conductive hearing loss, unilateral hearing loss, minimal sensorineural hearing loss (unaided), mild-to-moderate sensorineural hearing loss (aided), articulation disorders, speech and language delays, developmental delays, central auditory processing (CAP) disorder, children for whom English is a second language (ESL), and normal hearing but need to develop listening and academic skills. Teachers who present with voice disorders or chronic voice problems have also been shown to benefit from use of sound-field FM amplification. Refer to Chapters 4 and 8 for additional information on sound-field FM amplification research findings.

## Transmitter and Microphone Options

There are several options for wearing the transmitter/microphone: clip the transmitter to the neckloop, clip the transmitter to a belt or waistband, carry the transmitter in a pocket or belt pack, or attach the transmitter to a large belt worn across the body (see Chapter 9). The microphone must be positioned at a distance of 3–6 inches from the teacher's mouth in order to achieve maximum benefit from the FM system. Ideal placement of the microphone is on the upper mid-chest area. Microphone placement options include: clipping the microphone directly to clothing in the upper mid-chest area, attaching the microphone to the neckloop provided for a lavaliere style, or placing a small piece of fabric over a tie, shirt, or blouse before clipping on the microphone to prevent snags in clothing. An alternative is using a head boom microphone or assembly (shown in Figure 9–9 in Chapter 9) which keeps the microphone close to the teacher's mouth at all times. Interestingly, the majority of teachers in the ICA pilot project preferred to use the boom microphone.

The following suggestions are provided for classroom teachers to ensure successful use of the transmitter/microphone:

- Position the microphone 3–6 inches from your mouth. This allows a good signal to be picked up and is far enough from the mouth for attenuation of extraneous noises generated during speech. Use the windscreen over the microphone to help reduce additional noise. Lavaliere style microphones must be worn with a clip to prevent clothing noise.
- Speak in a natural voice. A normal conversational speech level will provide an adequate speech signal and it is not necessary to increase the intensity of your voice because the FM system is providing amplification to broadcast the signal.
- Avoid wearing any jewelry that may rub or hit against the microphone or microphone cord and produce unwanted stimuli for students.
- Use the ON/OFF control on the transmitter or receiver/ amplifier to mute the system when needed, such as during private conversations with a student, parent, or other classroom visitor. Since this equipment is capable of transmitting up to 300 feet, private conversations in the hallway may indeed be broadcast to the class if the transmitter is not turned off. Some sound-field FM systems do offer mute switch capability.
- Turn the system off when leaving the classroom and at the end of the day.
- Recharge the transmitter's battery each night. The operating time for the transmitter when using rechargeable batteries is approximately eight hours. A 12-hour use may be anticipated for alkaline disposable batteries. Do not recharge disposable batteries as they may rupture and damage the transmitter.
- Locate a safe place away from excessive heat, cold, and dampness for overnight storage of the transmitter.
- Avoid bumping or dropping the microphone/transmitter. Treat cords, microphone, and batteries gently. These are often the areas that malfunction first.
- Avoid winding the microphone connector cable around the unit. This could result in a break in the connecting cord which will produce static or noise in the FM system, or if a complete break occurs there will be no sound. If static

occurs, manipulate the microphone connector cord to determine if that might be the location of the problem.

- Avoid blowing into the microphone when testing it. Instead, rub your fingers over the windscreen or snap your fingers near the microphone. Conduct the Ling 5-Sound Test by saying "ah, oo, ee, s, sh." These speech sounds represent the full frequency range of speech.

- Use the adapter cord to connect the transmitter to auxiliary sound sources such as the VCR, CD ROM, computer, cassette tape player, etc. The adapter cord should be plugged into the auxiliary sound source jack on the teacher transmitter. The receiver/amplifier also has an auxiliary sound source input jack and this may be used as an alternative coupling method.

- Consult with the audiologist when volume control adjustments are needed. Also contact the audiologist if there are any problems with the transmitter/microphone or any other components of the FM system.

## SPEAKER OPTIONS

Placement of the speakers may be different for each classroom depending upon room configuration, furniture arrangement, and competing noise sources. Speaker position influences the overall quality of sound in the classroom. It may be necessary to try different locations until the desired amplified acoustic environment is achieved. Speakers may be secured to speaker stands, mounted with brackets on walls or furniture, or placed on shelves or other safe areas. Four speakers were used for the ICA project. A minimum of two to four speakers is currently recommended for most classroom settings in order that all students seated in the main teaching area will be within the critical listening distance. See Chapter 3 for a discussion of critical listening distance. A sound-field FM amplification system is now available with a ceiling mounted speaker arrangement with four main speakers and an attenuated center speaker. The system fits in the area of an acoustical ceiling tile and requires only one electrical connection. Other systems use wireless speakers and this is a benefit in that speaker wires do not have to be carefully secured within the classroom.

Speaker placement variables to be considered include distance, height, and concentration in the main teaching area. Specific details related to these variables are as follows:

- Distance between speakers should be approximately two-thirds the distance between the speakers and the primary listening area. When a greater separation between speakers is necessary, turning them slightly inward may result in a better sound effect. If the distance between speakers is too great, dead spots will occur in the classroom and if the distance is too small the stereo effect is reduced. The goal with speaker placement is for all students to be seated within the critical listening distance (Allen, 1991; Crandell, 1991b). See Chapter 3 for additional information.
- Speakers should be placed at or above ear height of the students. Approximate ear heights of seated students provided by Allen (1991) are a good reference to use.
- Place speakers in areas of the classroom where students receive the majority of class instruction in the main seating area. Speakers should not be placed directly across from each other for this can produce feedback. They should be placed with no furniture or barriers directly in front of them and avoid placing speakers flush with corner walls. After speaker placement has been determined, the speaker stands may be secured to protect the system. Speaker wires should be secured to avoid risk of accidental tripping over the wires.

## SETTING THE RECEIVER/AMPLIFIER VOLUME CONTROL

An audiologist should measure the ambient noise level in the classroom and the teacher's voice with a sound-level meter and assist the teacher in setting the volume control. Participants should now use the transmitter/microphone and follow the procedure described below. Encourage other participants to assist in judging the appropriate amplification level. After teachers have even a brief experience in using the system they rarely volunteer to give up using it in their classroom. Effective demonstration sells the concept and the product since the benefits are obvious to users:

- Turn on the receiver/amplifier and the transmitter/microphone and begin speaking.
- Continue talking and move around the room to ensure that there are no dead spots or over-amplification areas.
- Sit or stoop to the approximate ear level of the students.

- After examining these variables the most comfortable listening level for the classroom can be established. Set the volume control on the receiver/amplifier to the most comfortable level.
- The volume control setting is correct if the teacher is able to approach each speaker without causing feedback. If feedback occurs, the volume setting will need to be decreased or the speaker placement might be adjusted.
- When using an auxiliary input, connect the adapter cord from the external source into the auxiliary input jack on the receiver/ amplifier. Increase the auxiliary volume control to achieve a comfortable listening level.

## Maintenance and Troubleshooting Tips

It is necessary to become familiar with the care and maintenance information available in the equipment manual (see Appendixes B and C in Chapter 9). It is important to have available disposable alkaline batteries for emergencies such as a power outage during the charging period or forgetting to charge the system overnight. The audiologist or other support person in the district will install the system to ensure proper connection of speaker wires, placement of speakers, and setting the volume control. Speaker wires may be affixed with adhesive cable clips, concealed in wire strip covers or rubber door threshold covers, concealed above the acoustical tile ceiling, or suspended from the ceiling with metal clip mounts. Use of tape is an alternative for securing speaker wires; however, changes in classroom temperatures may require frequent replacement. The equipment manufacturer should provide adequate information regarding installation of speakers, options for daisy chain speaker installation, and other necessary instructions to ensure a properly functioning system.

Additional references on sound-field FM amplification systems: Allen (1990), Beaulac, Pehringer and Shough (1991), Beck and Nance (1991), Berg (1987, 1990, 1993), Blair (1991), Blake, Torpey, and Wertz (1987), Blake, Field, Foster, Plott, and Wertz (1991), Boothroyd (1992), Brandt (1989), Compton (1989, 1993), Crandell (1993c), Crandell and Smaldino (1992), Crandell, Smaldino, & Flexer (1994), Davis (1991, 1992), Evans (1992), Flexer (1992a, 1993), Florida Department of Education (1991b, 1994), Hammond (1991), Iowa Department of Education (1993), Leavitt (1991), Lewis (1994), Lewis, Feigin, Karasek, and Stelmachowicz (1991), Maxon (1992), Mills (1991), Ottring, Smaldino, Plakke, and Bozik (1992), Phonic Ear (1992, 1993a, 1993b), Pimentel (1988), Ross (1987, 1992), Sudler and Flexer (1986), and Worner (1988).

## SUMMARY AND CONCLUSIONS

Use of a sound-field FM amplification system provides the teacher with a unique opportunity to maximize the listening and learning opportunities in the classroom. In summary, the salient points identified through sound-field FM amplification research support benefits for both students and teachers. Among the research findings are:

- Improved academic achievement, especially for younger students
- Decreased distractibility and increased on-task behavior
- Increased attention to verbal instruction and activities and improved understanding
- Decreased number of requests for repetition
- Decreased frequency of need for verbal reinforcers to facilitate test performance
- Decreased test-taking time
- Improved spelling ability under degraded listening conditions
- Increased sentence recognition ability
- Improved listening test scores
- Increased language growth
- Improved student voicing when speaking
- Increased student length of utterance
- Increased confidence when speaking
- Increased preference by teachers and students for sound-field FM amplification in the classroom
- Improved ease of listening and teaching
- Reduced vocal strain and fatigue for teachers
- Increased mobility for teachers
- Reduced special education referral rate
- Increase in seating options for students with hearing loss
- Cost-effective means of enhancing the listening and learning environment.

These findings amplify the logic for using sound-field FM amplification. Research has shown that use of the sound-field FM equipment produced changes in each of the four levels (sensation, perception and processing, cognitive, response) in the listening process described earlier. **A very basic philosophy that supports use of sound-field FM amplification in the classroom is that "hearing is the primary channel for learning, and the more children can hear, the better they learn"** (Ross, 1991, p. 410).

Additional references on the benefits of sound-field FM amplification: Allen (1993), Allen and Patton (1990), Benafield (1990), Berg (1987), Berg, Bateman, and Viehweg (1989), Crandell (1991b, 1993c), Crandell and Bess, 1987b), Crandell and Smaldino (1992), Elliot (1982), Flexer (1989, 1992a, 1993), Flexer, Millin, and Brown (1990), Gilman and Danzer (1989), Jones, Berg, and Viehweg (1989), Nabelek and Nabelek (1985), Osborn, VonderEmbse, and Graves (1989), Ray (1988, 1989, 1992), Ray, Sarff, and Glassford (1984), Rosenberg, Blake-Rahter, Allen, and Redmond (1994) Ross (1991), Sarff (1981), Sarff, Ray, and Bagwell (1981), Schermer (1991), and Zabel and Tabor (1993).

## REFERENCES

Allen, L. (1990). WHAT? Speaker system helps in classroom. *Educational Audiology Association Newsletter, 7*(3), 7.

Allen, L. (1991). *A school handbook on classroom amplification equipment.* Elkader, IA: Keystone Area Education Agency.

Allen, L. (1993). Promoting the usefulness of classroom amplification. *Educational Audiology Monograph, 3,* 32–34.

Allen, L., & Patton, D. (1990). *Effects of sound field amplification on students' on-task behavior.* Paper presented at the American Speech, Language, and Hearing Association Convention, Seattle, WA.

Anderson, K. (1989a). *Screening instrument for targeting educational risk (SIFTER).* Austin, TX: PRO-ED.

Anderson, K. (1989b). Speech perception and the hard-of-hearing child. *Educational Audiology Monograph, 1*(1), 15–29.

Beaulac, D., Pehringer, J., & Shough, L. (1989). Listening devices: Available options. In C. Compton (Ed.), *Seminars in hearing: Assistive devices, 10*(1) (pp. 11–29). New York: Thieme Medical Publishers.

Beck, L., & Nance, G. (1989). Hearing aids, assistive devices, and telephones: Issues to consider. In C. Compton (Ed.), *Seminars in hearing: Assistive devices, 10*(1) (pp. 78–89). New York: Thieme Medical Publishers.

Benafield, N. (1990). The effects of sound field amplification on the attending behaviors of speech and language-delayed preschool children. Unpublished master's thesis, University of Arkansas at Little Rock.

Berg, F. (1986). Classroom acoustics and signal transmission. In F. Berg, J. Blair, J. Viehweg, & A. Wilson-Vlotman (Eds.), *Educational audiology for the hard of hearing child* (pp. 157–180). New York: Grune & Stratton.

Berg, F. (1987). *Facilitating classroom listening: A handbook for teachers of normal and hard of hearing children.* Boston: College-Hill Press.

Berg, F. (1990). Sound field amplification: Uses and applications. *Educational Audiology Association Newsletter, 7*(3), 6.

Berg, F. (1991). Historical perspectives of educational audiology. In C. Flexer (Ed.), *Seminars in hearing: Current audiologic issues in the educational management of children with hearing loss, 12*(4), 305–316. New York: Thieme Medical Publishers.

Berg, F. (1993). *Acoustics and sound systems in schools.* San Diego, CA: Singular Publishing Group, Inc.

Berg, F., Bateman, R., & Viehweg, S. (1989). *Sound field FM amplification in junior high school classrooms.* Paper presented at the American Speech-Language-Hearing Association Convention, St. Louis, MO.

Berry, V. (1988). Classroom intervention strategies and resource materials for the auditorily handicapped child. In R. Roeser & M. Downs (Eds.), *Auditory disorders in school children* (pp. 325–349). New York: Thieme-Stratton.

Bess, F., & Logan, S. (1984). Amplification in the educational setting. In J. Jerger (Ed.), *Pediatric audiology* (pp. 147–176). San Diego, CA: College Hill Press.

Bess, F., & McConnell, F. (1981). *Audiology, education and the hearing impaired child.* St. Louis, MO: C.V. Mosby.

Bess, F., Sinclair, J., & Riggs, D. (1984). Group amplification in schools for the hearing-impaired. *Ear and Hearing, 5,* 138–144.

Blair, J. (1977). Effects of amplification, speechreading, and classroom environment on reception of speech. *Volta Review, 79,* 443–449.

Blair, J. (1991). Educational audiology & methods for bringing about change in schools. In C. Flexer (Ed.), Seminars in hearing: *Current audiologic issues in the educational management of children with hearing loss, 12*(4) (pp. 318–328). New York: Thieme Medical Publishers.

Blake, R., Torpey, C., & Wertz, P. (1987). *Preliminary findings: Effect of FM auditory trainers on auditory attending behaviors of learning disabled children.* Minneapolis: Telex Communications.

Blake, R., Field, B., Foster, C., Plott, F., & Wertz, P. (1991). Effect of FM auditory trainers on attending behaviors of learning-disabled children. *Language, Speech and Hearing Services in the Schools, 22,* 111–114.

Bloomfield, R. (1987). Classroom noise control strategies. *Hearsay,* 41–44.

Boothroyd, A. (1992). The FM wireless link: An invisible microphone cable. In M. Ross (Ed.), *FM auditory training systems: Characteristics, selection and use* (pp. 1–19). Timonium, MD: York Press.

Borrild, K. (1978). Classroom acoustics. In M. Ross & T. Giolas (Eds.), *Auditory management of hearing impaired children* (pp. 145–179). Baltimore: University Park Press.

Brandt, F. (1989). Microphones and assistive listening devices: A tutorial. In C. Compton (Ed.), *Seminars in hearing: Assistive devices, 10*(1), (pp. 31–41). New York: Thieme Medical Publishers.

Butler, K. (1975). *Auditory perceptual skills: Their measurement and remediation with preschool and school-age children*. Paper presented at the American Speech-Language-Hearing Association Convention, Washington, D.C.

Cherow, E. (1986). *The ear and hearing*. Washington, DC: Gallaudet College Press.

Collins, S. (1989). *Sound hearing* (audiotape). Eugene, OR: Garlic Press.

Compton, C. (1989). *Assistive devices: Doorways to independence*. Washington, DC: Gallaudet University Press.

Compton, C. (1993). Why use assistive listening devices? *SHHH Journal, 14*(1), 14–17.

Conway, L. (1990). Issues relating to classroom management. In M. Ross (Ed.), *Hearing-impaired children in the mainstream* (pp. 131–157). Parkton, MD: York Press.

Crandell, C. (1991a). *An update on classroom acoustics for hearing impaired children*. Paper presented at the American Speech-Language-Hearing Association Convention, Atlanta, GA.

Crandell, C. (1991b). The effects of classroom amplification on children with normal hearing: Implications for intervention strategies. *Educational Audiology Monograph, 2* 18–38.

Crandell, C. (1992). Classroom acoustics for hearing-impaired children. *Journal of theAcoustical Society of America, 92*(4), 2470.

Crandell, C. (1993a). Speech recognition in noise by children with minimal degrees ofsensorineural hearing loss. *Ear & Hearing, 14*(3), 210–216.

Crandell, C. (1993b). A comparison of commercially-available frequency modulation soundfield amplification systems. *Educational Audiology Monograph, 3* , 15–20.

Crandell, C., & Bess, F. (1986). *Speech recognition of children in a "typical" classroom setting*. Paper presented at the American Speech-Language-Hearing Association Convention, Detroit, MI.

Crandell, C., & Bess, F. (1987a). *Developmental changes in speech recognition in noise and reverberation*. Paper presented at the American Speech-Language-Hearing Association Convention, New Orleans, LA.

Crandell, C., & Bess, F. (1987b). *Sound-field amplification in the classroom setting*. Paper presented at the American Speech-Language-Hearing Association Convention, New Orleans, LA.

Crandell, C., & Flexer, C. (1994). Improving classroom acoustics: A weekend with the experts seminar. Orlando, FL.

Crandell, C., & Smaldino, J. (1992). Sound field amplification in the classroom setting. *American Journal of Audiology, 1*(4), 14–16.

Crandell, C., Smaldino, J., & Flexer, C. (1994). *Theoretical and applied issues in sound field amplification*. Paper presented at the American Academy of Audiology annual convention, Richmond, VA.

Crum, M., & Matkin, N. (1976). Room acoustics: The forgotten variable. *Language, Speech & Hearing Services in the Schools, 3*, 106–110.

Davis, D. (1991). Utilizing amplification devices in the regular classroom. *Hearing Instruments, 42*(7), 18–23.

Davis, D. (1992). Sound field amplification: A look more cautiously. *Educational Audiology Association Newsletter, 9*(1), 8–9.

DeConde, D. (1984). Children with central auditory processing disorders. In R. Hull & K. Dilka (Eds.), *The hearing impaired child in school* (pp. 141–162). Orlando: Grune & Stratton.

Dirks, D., Morgan, D., & Dubno, J. (1982). A procedure for quantifying the effects of noise on speech recognition. *Journal of Speech & Hearing Disorders, 47*, 114–122.

Downs, D., & Crum, M. (1978). Processing demands during auditory learning under degraded listening conditions. *Journal of Speech & Hearing Research, 21*, 702–714.

Edwards, C. (1991a). Assessment and management of listening skills in school-aged children. In C. Flexer (Ed.), *Seminars in hearing: Current audiologic issues in the educational management of children with hearing loss, 12*(4) (pp. 389–401). New York: Thieme Medical Publications.

Edwards, C. (1991b). The transition from auditory training to holistic auditory management. *Educational Audiology Monograph, 2*(1), 1–17.

Edwards, C. (1993c). The changing classroom environment: Implications for auditory management. *Educational Audiology Newsletter, 10*(3), 8–9.

Elliott, L. (1979). Performance of children aged 9 to 17 on a test of speech intelligibility in noise using sentence material with controlled word predictability. *Journal of the Acoustical Society of America, 66*, 651–653.

Elliott, L. (1982). Effects of noise on perception of speech by children and certain handicapped individuals. *Sound and Vibration, December*, 9–14.

Evans, C. (1992). Troubleshooting FM systems. In M. Ross (Ed.). *FM auditory training systems: Characteristics, selection and use* (pp. 125–155). Timonium, MD: York Press.

Federal Communications Commission. (1992, April 7). Amendment of Part 15. (92-163), E.T. Docket No. 91-150. Additional frequencies for auditory assistance devices for the hearing impaired. Washington, DC.

Finitzo, T. (1988). Classroom acoustics. In R. Roeser & M. Downs (Eds.), *Auditory disorders in school children*, (2nd ed.) (pp. 221–233). New York: Thieme-Stratton.

Finitzo-Hieber, T., & Tillman, T. (1978). Room acoustic effects on monosyllabic word discrimination ability for normal and hearing impaired children. *Journal of Speech and Hearing Research, 21*, 440–448.

Fior, R. (1972). Psychological maturation of auditory function between 3 and 13 years of age. *Audiology, 11*, 317–321.

Fisher, L. (1976). *Fisher's auditory problems checklist.* Cedar Rapids, IA: Grant Wood Area Education Agency.

Flexer, C. (1989). Turn on sound: An odyssey of sound field amplification. *Educational Audiology Association Newsletter, 5*(5), 6–7.

Flexer, C. (1992). Classroom public address systems. In M. Ross (Ed.), *FM auditory training systems: Characteristics, selection and use* (pp. 189–209). Timonium, MD: York Press.

Flexer, C. (1993). *Decisions in the selection and management of classroom amplification systems.* Paper presented at Pediatric Audiology Update, St. Petersburg, FL.

Flexer, C., Millin, J., & Brown, L. (1990). Children with developmental disabilities: The effect of sound field amplification on word identification. Language, *Speech and Hearing Services in Schools, 21,* 177–182.

Flexer, C., Richards, C., & Buie, C. (1993). *Sound-field amplification for regular kindergarten and first grade classrooms: A longitudinal study of fluctuating hearing loss and pupil performance.* Paper presented at the American Academy of Audiology Convention, Phoenix, AZ.

Florida Commission on Education Reform and Accountability. (1992). *Blueprint 2000: A system of school improvement and accountability.* Tallahassee, FL.

Florida Department of Education. (1988). *A resource manual for the development and evaluation of special programs for exceptional students: Volume II-D: Deaf and hard-of-hearing.* Tallahassee, FL.

Florida Department of Education. (1991a). *Meeting the educational needs of students who have hearing losses: Module 1: Hearing loss and audiogram interpretation.* Tallahassee, FL.

Florida Department of Education. (1991b). *Meeting the educational needs of students who have hearing losses: Module 2: Amplification systems.* Tallahassee, FL.

Florida Department of Education. (1991c). *Meeting the educational needs of students who have hearing losses: Module 3: Speech and language development of students with hearing losses.* Tallahassee, FL.

Florida Department of Education. (1991d). *Meeting the educational needs of students who have hearing losses: Module 4: Classroom and instructional concerns and modifications.* Tallahassee, FL.

Florida Department of Education. (1994). *Improving classroom acoustics: Inservice training manual.* Tallahassee, FL.

Gengel, R. (1971). Room acoustics effects on monosyllabic word discrimination ability for normal and hearing-impaired children. *Journal of Auditory Research, 11,* 219–222.

Gilman, L., & Danzer, V. (1989). *Use of FM sound field amplification in regular classrooms.* Paper presented at the American Speech-Language-Hearing Association Convention, St. Louis, MO.

Hammond, L. (1991). *FM auditory trainers: A winning choice for students, teachers & parents.* Minneapolis, MN: Gopher State Litho Corp.

Hart, P. (1983). Classroom acoustical environments for children with central auditory processing disorders. In E. Lasky & J. Katz (Eds.), *Central auditory processing disorders: Problems of speech, language, and learning* (pp. 343–352). Baltimore, MD: University Park Press.

Harris, C. (1979). *Handbook of noise control.* New York: McGraw-Hill.

Holzhauser-Peters, L., & Husemann, D. (1988). Classroom observation: Getting the complete picture. *The Clinical Connection, 2*(4), 16–19.

Iowa Department of Education. (1993). *Iowa resource manual for the education of students with hearing loss and educational audiology.* Des Moines, IA.

Jones, J., Berg, F., & Viehweg. (1989). Listening of kindergarten students under close, distant, and sound field FM amplification conditions. *Educational Audiology Monograph, 1*(1), 56–65.

Keith, R. (1988). Tests of central auditory function. In R. Roeser & M. Downs (Eds.), *Auditory disorders in school children* (2nd ed.) (pp. 81–97). New York: Thieme-Stratton.

Kent, R. (1992). Auditory processing of speech. In J. Katz, N. Stecker, & D. Henderson (Eds.), *Central auditory processing: A transdisciplinary view* (pp. 93–105). St. Louis: Mosby-Year Book, Inc.

Lasky, E. (1983). Parameters affecting auditory processing. In E. Lasky &. J. Katz (Eds.), *Central auditory processing disorders: Problems of speech, language, and learning* (pp. 11–29). Baltimore: University Park Press.

Lasky, E., & Chapandy, A. (1976). Factors affecting language comprehension. *Language, Speech and Hearing Services in Schools, 7,* 159–168.

Lasky, E., & Cox, L. (1983). Auditory processing and language interaction: Evaluation and intervention strategies. In E. Lasky &. J. Katz (Eds.), *Central auditory processing disorders: Problems of speech, language, learning* (pp. 243–268). Baltimore: University Park Press.

Lasky, E., & Tobin, H. (1973). Linguistic and non-linguistic competing message effects. *Journal of Learning Disabilities, 6,* 243–250.

Leavitt, H. (1978). The acoustics of speech production. In M. Ross & T. Giolas (Eds.), *Auditory management of hearing-impaired children: Principles and prerequisites for intervention.* Baltimore: University Park Press.

Leavitt, R. (1991). Group amplification systems for students with hearing impairment. In C. Flexer (Ed.), *Seminars in hearing; Current audiologic issues in the educational management of children with hearing loss, 12*(4), 380–388. New York: Thieme Medical Publishers.

Leavitt, R., & Flexer, C. (1991). Speech degradation as measured by the rapid speechtransmission index (RASTI). *Ear & Hearing, 12,* 115–118.

Leggett, S., Brubaker, C., Cohodes, A., & Shapiro, A. (Eds.). (1977). *Planning flexible learning places.* New York: McGraw-Hill.

Lewis, D. (1994). Assistive devices for classroom listening. *American Journal of Audiology, 3*(1), 58–69.

Lewis, D., Feigin, J., Karasek, A., & Stelmachowicz, P. (1991). Evaluation and assessment of FM systems. *Ear and Hearing, 12,* 268–280.

Madell, J. (1990). Managing classroom amplification. In M. Ross (Ed.). *Hearing-impaired children in the mainstream* (pp. 95–118). Parkton, MD: York Press.

Markides, A. (1986). Speech levels and speech-to-noise ratios. *British Journal of Audiology, 20,* 115–120.

Maxon, A. (1992). FM selection and use for school-age children. In M. Ross (Ed.), *FM auditory training systems: Characteristics, selection, and use* (pp. 103–124). Timonium, MD: York Press.

McCroskey, F., & Devens, J. (1975). Acoustic characteristics of public school classrooms constructed between 1890 and 1960. *NOISEXPO Proceedings,* 101–103.

Mills, M. (1991). A practical look at classroom amplification. *Educational Audiology Monograph, 2*(1), 39–42.

Nabelek, A., & Nabelek, I. (1985). Room acoustics and speech perception. In J. Katz (Ed.), *Handbook of clinical audiology* (3rd ed.) (pp. 834–846). Baltimore: Williams & Wilkins.

Nabelek, A., & Pickett, J. (1974). Monaural and binaural speech perception through hearing aids under noise and reverberation with normal and hearing-impaired listeners. *Journal of Speech and Hearing Research, 17,* 724–739.

Nabelek, A., & Robinson, P. (1982). Monaural and binaural speech perception in reverberation for listeners of various ages. *Journal of the Acoustical Society of America, 71*(5), 1242–1248.

National Institute on Deafness and Other Communication Disorders. (1990). *I love what I hear!* Bethesda, MD: National Institutes of Health.

Neuman, A., & Hochberg, I. (1983). Children's perception of speech in reverberation. *Journal of the Acoustical Society of America, 73*(6), 2145–2149.

Neuss, D., Blair, J., & Viehweg, S. (1991). Sound field amplification: Does it improve word recognition in a background of noise for students with minimal hearing impairments? *Educational Audiology Monograph, 2*(1), 43–52.

Northern, J., & Downs, M. (1991). *Hearing in children* (4th ed.). Baltimore: Williams & Wilkins.

Olsen, W. (1981). The effects of noise and reverberation on speech intelligibility. In F. Bess, B. Freeman & J. Sinclair (Eds.). *Amplification in education* (pp. 151–163). Washington, DC: A.G. Bell Association for the Deaf.

Osborn, J., VonderEmbse, D., & Graves, L. (1989). Development of a model program using sound field amplification for prevention of auditory-based learning disabilities. Unpublished Study, Putnam County Office of Education, Ottawa, OH.

Ottring, S., Smaldino, J., Plakke, B., & Bozik, M. (1992). *Comparison of two methods of improving classroom S/N ratio.* Paper presented at the American Speech, Language, and Hearing Association Convention, San Antonio, TX.

Papso, C., & Blood, I. (1989). Word recognition skills of children and adults in background noise. *Ear & Hearing, 10*(4), 235–236.

Peters, T., & Austin, N. (1985). *A passion for excellence.* New York: Warner.

Phonic Ear, Inc. (1992). *Easy listener free field sound system: Installation tips.* Petaluma, CA.

Phonic Ear, Inc. (1993a). *All about FM.* Petaluma, CA.

Phonic Ear, Inc. (1993b). *Easy listener free field sound system: User instructions.* Petaluma, CA.

Pimentel, R. (1988). Classroom amplification systems for the partially hearing student. In R. Roeser & M. Downs (Eds.), *Auditory disorders in school children* (2nd ed.) (pp. 234–245). New York: Thieme-Stratton.

Rabbit, P. (1966). Recognition: Memory for words correctly heard in noise. *Psychonomic Sciences, 6,* 383–384.

Ray, H. (1987). Put a microphone on the teacher: A simple solution for the difficult problems of mild hearing loss. *The Clinical Connection, Spring,* 14–15.

Ray, H. (1988). *Mainstream amplification resource room study (MARRS): A national diffusion network project.* Norris City, IL: Wabash & Ohio Valley Special Education District.

Ray, H. (1989). Project MARRS-an update. *Educational Audiology Association Newsletter, 5*(5), 4–5.

Ray, H. (1992). *Summary of MARRS adoption data validated in 1992.* Norris City, IL: Wabash & Ohio Valley Special Education District.

Ray, H., Sarff, L., & Glassford, F. (1984). Soundfield amplification: An innovative educational intervention for mainstreamed learning disabled students. *The Directive Teacher, 6*(2), 18–20.

Richards, C., Flexer, C., Brandy, W., & Wray, D. (1993). Signal-to-noise enhancing devices can improve kids reading skills. *Hearing Instruments, 44*(1), 12–15.

Rosenberg, G., Blake-Rahter, P., Allen, L., & Redmond, B. (1994). Improving classroom acoustics: A multi-district pilot study on FM classroom amplification. Poster session presented at the American Academy of Audiology annual convention, Richmond, VA.

Ross, M. (1978). Classroom acoustics and speech intelligibility. In J. Katz (Ed.), *Handbook of clinical audiology* (pp. 469–478). Baltimore: Williams & Wilkins.

Ross, M. (1987). FM auditory training systems as an educational tool. *Hearing Rehabilitation Quarterly, 12*(4), 4–6.

Ross, M. (1991). A future challenge: Educating the educators and public about hearing loss. In C. Flexer (Ed.), *Seminars in hearing: Current audiologic issues in the educational management of children with hearing loss, 12*(4), (pp. 402–13). New York: Thieme Medical Publishers.

Ross, M. (1992). Room acoustics and speech perception. In M. Ross (Ed.), *FM auditory training systems: Characteristics, selection and use* (pp. 21–43). Timonium, MD: York Press.

Ross, M., & Giolas, T. (1971). Effects of three classroom listening conditions on speech intelligibility. *American Annals of the Deaf, 116*(6), 580–584.

Ross, M., Maxon, A., & Brackett, D. (1982). *Hard of hearing children in regular schools.* Englewood Cliffs, NJ: Prentice-Hall.

Rupp, R. (1978). The audiologist's role in the evaluation of auditory perceptual and processing abilities in young school-age children. *Audiology, 3,* 7.

Rupp, R., Jackson, P., & McGill, N. (1986). Listening problems = academic distress. *Hearing Instruments, 37*(9), 20–24.

Sanders, D. (1965). Noise conditions in normal school classrooms. *Exceptional Children, 31,* 34–353.

Sanders, D. (1985). *Hearing impairment, amplification and learning.* Paper presented at Florida Educators of the Hearing Impaired Convention, St. Petersburg, FL.

Sanger, D. (1988). Observational profile of classroom communication. *Clinical Connection, 2*(2), 11–13.

Sanger, D., Freed, J., & Decker, T. (1985). Behavioral profile of preschool children suspected of auditory language processing problems. *The Hearing Journal, 10,* 17–20.

Sarff, L. (1981). An innovative use of free-field amplification in classrooms. In R. Roeser & M. Downs (Eds.), *Auditory disorders in school children* (pp. 263–272). New York: Thieme-Stratton.

Sarff, L., Ray, H., & Bagwell, C. (1981). Why not amplification in every classroom? *Hearing Aid Journal, 34*(12), 11.

# 182 Sound-Field FM Amplification: Theory and Practical Applications

Schermer, D. (1991). Briggs sound amplified classroom study. Unpublished study, Briggs Elementary School, Maquoketa, IA.

Schneider, D. (1992). Audiologic management of central auditory processing disorders. In J. Katz, N. Stecker, & D. Henderson (Eds.), *Central auditory processing: A transdisciplinary view* (pp. 161–176). St. Louis: Mosby-Year Book, Inc.

Silverstone, D. (1982). Considerations for listening and noise distraction. In P. Sleeman & D. Rockwell (Eds.), *Designing learning environments* (p. 79). New York: Longman.

Smoski, W. (1990). Use of CHAPPS in a children's audiology clinic. *Ear and Hearing, 11* (Suppl.), 53S–56S.

Smoski, W., Brunt, M., & Tannahill, J. (1992). Listening characteristics of children with central auditory processing disorders. *Language, Speech and Hearing Services in Schools, 23,* 145–152.

Truesdale, S. (1990). Whole-body listening: Developing active auditory skills. *Language, Speech, and Hearing Services in Schools, 21,* 183–184.

Willeford, J., & Burleigh, J. (1985). *Handbook of central auditory processing disorders in children.* Orlando: Grune & Stratton.

Worner, W. (1988). An inexpensive group FM amplification system for the classroom. *The Volta Review, 90,* 29–36.

Yacullo, W., & Hawkins, D. (1987). Speech recognition in noise and reverberation by school age children. *Audiology, 26,* 235–246.

Zabel, H., & Tabor, M. (1993). Effects of classroom amplification on spelling performance of elementary school children. *Educational Audiology Monograph, 3,* 5–9.

# APPENDIX A
# STUDENT AND CLASSROOM OBSERVATION CHECKLISTS

Checklist of Classroom Observations for Children With Possible Auditory Processing Disorders (Sanger, Stick, & Smith, 1985).

Children's Auditory Processing Performance Scale (CHAPPS) (Smoski, 1990).

Classroom Observation Checklist for Use by SLP (Holzhauser-Peters & Husemann, 1988).

Evaluation of Children with Suspected Listening Difficulties (Edwards, 1991a).

Evaluation of Classroom Listening Behaviors (ECLB) (VanDyke, 1985; adaptation for ICA Project, Florida Department of Education, 1994; Rosenberg, Blake-Rahter, Allen, & Redmond, 1994).

Fisher's Auditory Problems Checklist (Fisher, 1976).

Listening and Learning Observation (LLO) (Florida Department of Education, 1994; Rosenberg, Blake-Rahter, Allen, & Redmond, 1994)

Observational Profile of Classroom Communication (Sanger, 1988).

SIFTER (Anderson, 1989a).

Willeford and Burleigh Behavior Rating Scale for Central Auditory Disorders (Willeford & Burleigh, 1985).

# APPENDIX B
# CLASSROOM DESCRIPTION WORKSHEET

The Classroom Description Worksheet was developed for use in the IMPROVING CLASSROOM ACOUSTICS (ICA) project (Florida Department of Education, 1994). The worksheet was devised to obtain information from the teacher about classroom acoustics and to record noise measurement data for the project. The following is an outline of items included on the worksheet.

## I. General Information
   A. Location of School in the Community
      (e.g., intersection of busy streets, quiet neighborhood, near a construction site, near airport, etc.)
   B. General Description of School Layout
      1. Exterior Structure
         (e.g., brick, concrete block, wood, etc.)
      2. Type of classrooms
         (e.g., self-contained, open, portables)
      3. Configuration
         (e.g., long hallways, enclosed building, description of common areas, etc.)
      4. Windows
         (e.g., size and placement, facing hallways, etc.)
      5. Type of Lighting
         (e.g., fluorescent or other)
      6. Landscaping
         (e.g., trees, shrubbery, earth mounds, etc.)
      7. Other Unique Features
         (e.g., detached exterior walls, carpeted exterior hallways, noise reduction signage, etc.)

## II. Classroom Features and Arrangement
   A. Grade Level, Number of Students, Room Dimensions
   B. Location of Classroom at School Site
      (Include age of classroom if different than original building.)
   C. Ceiling
      1. Height
      2. Acoustical Tile on Solid
      3. Acoustical Tile Suspended
      4. Plaster Finish
      5. Other Surface(s)

D. Floor
    1. Carpet (pad/no pad)
    2. Tile (rubber/other)
    3. Linoleum
    4. Terrazzo
    5. Wood
    6. Other Surface(s)

E. Walls
    1. Brick
    2. Concrete Block (painted/unpainted)
    3. Plaster
    4. Wood/Wood Paneling
    5. Partially Carpeted
    6. Draperies
    7. Chalkboards or Markerboard
    8. Other Surface(s)

F. Doors
    1. Metal/Wood
    2. Number/None
    3. Solid/Hollow Core
    4. Noise Lock Seal/Treatment
    5. Open/Closed Vent in Door
    6. Other Surface(s)/Feature(s)

G. Windows
    1. Size & Position
    2. Regular/Double Pane
    3. Number/None
    4. Draperies (regular/insulated)
    5. Blinds
    6. Other Feature(s)

H. Arrangement of Furniture
(include on classroom diagram; note if room is crowded; seating arrangement; presence of bookcases, study carrels, or partitions; location of chalk/markerboards; acoustical treatment of furniture, etc.)

I. Unique Classroom Features
(e.g., special purpose areas, etc.)

**III. Noise Sources**
(Noise sources are rated on a scale of 1 [quiet] to 5 [very noisy]).

A. Interior Noise Sources
    1. Bathroom in Classroom
    2. HVAC
    3. Fans (pedestal/ceiling)

      4.    Lighting (list type)
      5.    Computer
      6.    Other Interior Classroom Noise Sources
  B. External or Adjacent
    1.  Cafeteria
    2.  Playground
    3.  Courtyard
    4.  Street Noise
    5.  Gymnasium
    6.  Music/Band Room
    7.  Maintenance/Mechanical Room
    8.  Restroom/Drinking Fountain Area
    9.  Hallway(s)
    10.  Adjacent Classroom Noise
    11.  Building & Grounds Maintenance
    12.  Other External or Adjacent Noise Sources
  C. Unique or Unusual Noise Sources
(e.g., classroom near computer terminal & transmitting occurs 2 hours during the morning). Note the source(s) and length of exposure per day/week.

## IV. Acoustical Modifications or Other Noise Absorption Features in the Classroom
(e.g., bookcases, feltboards, acoustic partitions, etc.)

## V. Teacher Specific Information
  A. Class Description
    (e.g., generally quiet/noisy, relaxed/fast paced environment, multiple learning centers, etc.)
  B. Teacher's Style in Classroom
    (e.g., moving around, facing children when speaking, working with small groups, mode of transition from one activity to the next, use of gestures and visual aids, etc.)
  C. Teacher's Appraisal of His/Her Voice and Speech
    1.  Vocal intensity (soft, average, loud)
    2.  Vocal pitch (low, medium, high)
    3.  Vocal modulation
    4.  Rate of speech
    5.  Other features
  D. Teacher's Description of How Noise Sources Change
    (e.g., during the day/week, type of activity, etc.)

## VI. Classroom Arrangement Diagram
A worksheet is provided for teachers to draw a diagram of their classroom.

## VII. Noise Measurement Worksheet

The Noise Measurement Worksheet is part of the Classroom Description Worksheet devised for use in the ICA project. This form is used for recording noise measurements in both control and experimental classrooms. Complete directions accompany the noise measurement portion of the worksheet.

A. Classroom Grid

Identify the five classroom positions where noise measurements are taken.

B. Noise Measurement Data

Record all dBA and dBC noise measurements taken under unoccupied, occupied, and any special conditions (e.g., specific noisy times during the instructional day).

C. Teacher's Voice Measurement

Record unamplified and amplified measurements with the sound level meter held at 6 inches from the teacher's mouth.

# APPENDIX C
# CHECKLIST FOR RECOMMENDING, PURCHASING, INSTALLING, AND USING SOUND-FIELD FM AMPLIFICATION SYSTEMS IN CLASSROOMS

## School Location and Classroom Facilities
- Location of school within the community in regard to FM or VHF interference. (Manufacturers will provide information and in some cases lend scanner equipment to assist in determining which FM and VHF frequencies can be effectively used in certain locales. High-powered pagers, beepers, cellular phones, ham radios, and CB radios may cause interference.)
- Classroom style: self-contained, open plan, portable.
- Facility limitations (e.g., availability and condition of electrical outlets, limitations on mounting speakers and speaker wires, classroom traffic flow re speaker wire placement, ceiling height, room configuration features, etc.).

## Classroom Usage and Teaching Style
- Traditional or non-traditional classroom arrangement.
- Main teaching area; learning centers; other unique classroom features.
- Combination classes; team teaching.
- Teacher willingness to use system.

## Frequency Specifications
- Number of classes to be amplified; plan for growth. (Most systems have a 200-300 foot transmission range. Some of the VHF systems have a shorter transmission range.)
- Proximity of amplified classrooms.
- VHF vs FM frequencies; narrow band vs wide band frequencies.
- Number and type of personal FM systems used in the school; compatibility with sound-field FM amplification system.

## Equipment Needs and Options
- Portable vs stationary system.
- Wireless vs wired system.
- Microphone options: lavaliere, lapel, boom, collar.
- Number of speakers. (All students should be within the critical listening distance. A minimum of three speakers is recommended and preferably four to achieve maximum benefit for all students.)
- Muting capability. (Some systems offer a mute switch on the transmitter and others require that the transmitter/microphone be turned off.)

- Compatibility with district-owned personal FM systems.
- Auxiliary jacks, cables, and adapters for VCRs, CD ROM players, tape players, computer, etc. (These options should be available to enhance external audio sources rather than having to increase the volume to the point of distortion so that all students may hear.)
- Adapter for connection to student's personal FM systems.
- General ease of use.
- Installation tools for wired system (needle-nose pliers, cable wire clips, electrical tape, ceiling mount hooks, cable strip covers, doorway threshold covers, etc.).
- Maintenance costs (service contracts, disposable batteries for emergencies, replacement speaker wire, etc.)

### Equipment Manufacturer Considerations
- Manufacturer's support with installation (on-site vs telephone assistance; comprehensive installation, use, and troubleshooting manual; installation and training video, etc.).
- Availability of trial period.
- Length of warranty.
- Competitive pricing (estimated cost range: $300–1,765); length of time purchase price is in effect.
- Service availability (service turnaround time; loaner units; complimentary express mail return service; service contracts, etc.)

### Selection Capabilities Criteria
- Ease of installation. (Wired system should take approximately 30 minutes to install if the installation is well planned.)
- Transmission frequencies available.
- Frequency interference.
- Acoustic feedback.
- Number and position of speakers.
- Quality of amplified sound.
- Trial period. (Encourage observation by other stakeholders such as parents, PTO president, etc.)
- Durability.
- Portability.
- Cost and cost effectiveness.
- Qualitative and quantitative system capability evaluation. (Include observation and evaluative data from teachers, students, parents, and the school leadership team.)

### Cost Effectiveness
- Divide the cost of the unit by the total number of children and teachers who will benefit from the unit. (Include a system longevity factor to enhance the derived cost effectiveness figure.)

- Reliability or track record of the system.
- References from other school districts using the system.

## Administrative Support
- Provide information about known success of sound-field FM amplification in classrooms.
- Provide demonstration for administrators, PTO, faculty meetings, etc. (Hearing is believing!)
- Arrange for trial period in a selected classroom; provide inservice training and on-site support, troubleshooting instruction, etc.; measure student and teacher opinions to evaluate effectiveness.
- Provide inservice training on general noise control in schools, listening strategies, etc. to assist with improving the listening and learning environment.

## Funding Options
- Local district discretionary funds, technology grant funds, etc.
- 504 funding or special education funding for eligible students; everyone benefits.
- PTOs and service organizations.
- Grants from local foundations.
- Local business partners.

## Continuous Maintenance
- Role of the audiologist (district employee or contract audiologist) regarding initial consultation, provision of inservice training, etc.
- Assistance from school or district level media/telecommunications personnel for installation and maintenance assistance.
- Annual service contract or other maintenance agreement.
- Incidental costs (disposable batteries for emergencies, cable wire clips, replacement speaker wire, auxiliary input cables and adapters, etc.).
- Support materials (comprehensive equipment manuals, installation and use videos, etc.).

# CHAPTER
# 11

# LISTENING STRATEGIES FOR TEACHERS AND STUDENTS

*Carolyn Edwards*

**W**ith every new amplification option, there is the technical element and the human element. The introduction of sound-field FM systems into the classroom provides a new technology to enhance the listening environment. The introduction of new and thoughtful teaching strategies to improve the listening environment and students' listening abilities can maximize the use and benefit of the technology. Listening is the primary avenue for teaching and peer learning in the classroom. Optimal listening skills develop in an atmosphere that supports the enjoyment of sound, communication, storytelling, and experimentation. An ideal listening environment includes activities such as:

- sharing ideas and feelings
- listening centers where children can listen to music and taped stories
- daily storytelling by teacher and children
- children creating their own theater/plays
- singing and/or playing simple musical instruments
- creative experiments with sound such as creating new sound effects for a ghost story
- science experiments with sound such as the effects of different sounds on plant growth.

"Listening" in its broadest context is responding to, organizing, interpreting and evaluating sound in order to create meaning (Early Childhood Curriculum Committee, 1978). In other words, listening is the ability to detect, discriminate, identify, and comprehend various auditory signals. *Detection* is the ability to respond to sound, pay attention to sound, and learn not to respond when there is no sound. *Discrimination* is the ability to attend to differences among sounds, or to respond differently to different sounds. *Identification* is the ability to name or identify the sound heard. *Comprehension* is the ability to answer questions, follow instructions, paraphrase, or participate in a conversation (Erber, 1977).

The sense of hearing is used for several purposes: to comprehend the speech of others, to monitor our own speech, and to monitor the surrounding environment (Edwards, 1991). The focus of listening activities in early childhood curricula, however, often consists of comprehension of speech only. There is little or no discussion of strategies to enhance the environment in which students listen, despite the evidence that noise and reverberation have a deleterious effect on children's comprehension (Finitzo-Hieber & Tillman, 1978; Neuman & Hochberg, 1983; Yacullo & Hawkins, 1987). This chapter will focus on the ways in which teachers and students can increase their awareness of the listening environment, change the listening environment, and enhance speaker-listener skills when faced with more difficult listening conditions.

## MAKING SOUND VISIBLE

Development of listening skills in the classroom starts first with awareness of sounds that exist within the classroom environment. The first step is to make sound, and specifically noise, "visible" in the classroom. Although classroom noise is obviously interfering, teachers and students often adapt and accept noise as a normal characteristic of the classroom environment. In a study completed in schools in Quebec, Canada, 54 percent of classroom teachers and 77 percent of physical education teachers reported that noise usually caused communication problems in their respective work environments, in contrast to only 9 percent of office workers interviewed (Hétu, Truchon-Gagnon, & Bilodeau, 1990). Despite such data, there are no consistent efforts to reduce noise levels in the classroom. There are recommended guidelines, but no enforced acoustical standards for the design of new classrooms in the building codes with the exception of maximum allowable noise levels to prevent noise-induced hearing loss. This is clear evidence that the presence of noise has attracted little attention in educational circles.

Placement of a sound-field FM system in the classroom is often the first opportunity that teachers and students have to experience an improvement in speech-to-noise ratio. The change in the environment created by the new technology can provide the impetus for enhancing awareness of noise in the classroom. Once the sound-field FM system has been installed for about two to three weeks, ask the teacher turn the system off for part or all of one day. Have the children record their observations and consider questions such as those found in Table 11–1.

The contrast between no amplification and amplification is an excellent experience to heighten students' awareness of typical speech and noise levels in the classroom. For example, graduate students in speech-language pathology and audiology were completing a one-week course in educational audiology taught by the author in a room they had used as a lecture hall for the past two years. A sound-field FM system was installed for one week to provide them with first-hand experience of room amplification. At the end of this week, when the system was turned off during a discussion period, the students complained about the poor room acoustics, a factor that had previously never attracted their attention.

**Table 11–1.** Observations and questions for students and children.

| Questions to Ask |
| --- |

**For the Students**
- Are there any differences in the sound, and if so, what are they?
- Are there any noises that they notice that they hadn't noticed before? What are they?
- Is the noise level in the classroom softer or louder than before?
- Are they having difficulty hearing the teacher?
- Can they remember if they had difficulty hearing the teacher before the system was installed?

**For the Teacher**
- Do they use a louder voice when the system is off?
- How do the children respond to instructions?
- Do the teachers have to repeat more often?
- Is there a longer transition time between activities when the system is off?
- What is the background noise level in the classroom now?

Using a sound-level meter is another way to increase awareness of noise. Several electronics distributors sell inexpensive sound-level meters that can be purchased by the school. A discussion of sound-level meters is found in Chapter 5. Students in third grade or higher can easily learn to use the sound-level meter to measure sound in the classroom. Prior to measurement, it is instructive to have students estimate the loudness of the sound in decibels with a simple rating scale such as quiet, average, a little loud, and very loud, so that they begin to develop a comparative sense of sound intensities. Later, teachers can ask the students to rate the noise levels based on the degree of interference with communication and work caused by the noise, using a simple scale such as quiet, slightly interfering, moderately interfering, extremely interfering, and too loud to communicate or work independently. Students are often surprised to discover the intensity level of sounds such as the fans from the ventilation system, or the recess bell.

Encourage students to measure any sounds of interest occurring in the classroom, and notice if students become more aware of room noises as a result of the ongoing measurement of sounds. Teachers also can ask the students to measure sound levels in other parts of the school and report the results to other classes or teachers. Such reports can create an opportunity for inservice training of others. Audio-taping the classroom at different times of the day also can provide the students with more information about speech and noise levels in the class.

Teachers can experiment with a "silent" half day or day when no student sound is permitted. The only allowable communication is through gesture, drawing, or writing, and students are asked to keep all classroom sounds to a minimum. Paradoxically, it is often only when noise is reduced or eliminated that students and teachers become aware of its presence in the room. Usually after the experience of a silent day students are more aware of noises that exist in their classroom, and more specifically, the noises they themselves generate.

The sounds of chairs scraping on uncarpeted floors is also often ignored by teachers and students in the classroom. Ask each student to bring in two pairs of socks or tennis balls that have been cut in half, and attach the socks or balls to the bottom of the each student's chair legs to reduce the noise generated by chair movement. After two weeks remove the socks or balls and have the children report their observations. Did they notice a difference? How loud is the sound of the chairs moving? Can they understand when other students are talking and someone moves a chair at the same time? Again it is important to note that the auditory experience is heightened by experimenting first with the absence of noise, and then returning to the original noise levels.

It is also important for students to be aware of the differences in the speech levels of the teacher and their classmates as a function of distance and intensity levels. Some students may have difficulty hearing the classroom discussion adequately when they are sitting at a distance from the speaker or when they are listening to students who do not project their voices well (see Chapters 3 and 5). Passing around the transmitter of the sound-field FM system during class discussion often permits some students to be heard and understood for the first time. If students are unaware of differences in intensity in their voices, measure the loudness of other students' voices through the speakers, using the sound-level meter. One can never overdo awareness exercises. To maintain the "visibility of sound," it is useful to repeat activities such as the ones described here throughout the school year.

## CHANGING THE ENVIRONMENT

When teachers and students have a greater awareness of noise or poor quality speech signals, they are ready to initiate changes in the listening environment. After using the sound-field FM system for a few weeks, encourage daily monitoring of appropriate use by the students. For example, have the teacher start the day without the system turned on and wait for the students to notice. The teacher may do this from time to time and the student who first notices that the system is turned off gets rewarded. When the students pass the transmitter around during class discussions, ensure that the student does not begin to talk before receiving the transmitter. In classrooms using row seating, students in the seats at the back of the classroom are often the first to notice a poor signal during large group discussion where the microphone cannot easily be passed among the children. Encouraging students to say "I can't hear you" sets up a more proactive attitude toward listening expectations when students are sharing ideas.

When the sound-field system is not in use, the teacher might assign a "voice monitor" for the day. The monitor's responsibility would be to ask the speaker of the moment to talk louder, more clearly, or both. Although the sound-field FM system enhances the speech signal, monitoring and decreasing noise levels is a function best suited to people. The difficulty to date in any technological means of "removing noise" is that noise is defined by the person, not by a set of acoustical parameters (Boothroyd, 1994). For example, what is considered noise from several students working together at an adjacent table to the student reading to himself, is the primary speech signal for the students in that group. "Noise" is any speech or environmental sound that interferes with the individual's ability to focus on the task at hand.

Monitoring noise levels usually has been the prerogative of the classroom teacher. However, since students create the majority of the noise within a classroom it is appropriate for them to share responsibility for noise control with the teacher (Melancon, Truchon-Gagnon, & Hodgson 1990). In the author's experience, assigning students responsibility for noise control has worked very effectively in a number of classrooms. The power of a student monitor to ask classmates to be quiet is quite attractive to most students. Through their earlier experience in measuring sound levels with the sound-level meters, the students should be able to maintain some consistency in judgments of excessively loud sound. At a more sophisticated level, students in some of the higher grades might choose to form a committee to evaluate noise in the classroom and then to report back to the class with a set of recommendations for noise control. A hearing conservation program could also add some complementary information, particularly for young teenagers at risk for noise-induced hearing loss. The NIDCD Information Clearinghouse offers a free pamphlet for school children on hearing conservation.

Reserving an area of the classroom for quiet times also reinforces the teacher's support for noise reduction. Teachers have often used old bathtubs or a mound of pillows to create a "castle," or cocoon-like area, where the students can go to read or work on their own, without talking. In one classroom, the quiet area called "the office" was in such demand that there was a waiting line to get in. **Giving students themselves opportunities to make changes in their listening environments can create advocates for better acoustics.**

## SPEAKER-LISTENER SKILLS

Use of the sound-field FM system is an excellent way to teach speaker and listener skills. Good microphone technique is the first speaker skill that teachers and students may practice with FM systems. The teacher can experiment with various locations of the lapel or boom microphone, with respect to clarity and intensity of her or his speech. Involving the children in the judgments of optimal placement of the microphone gives them an opportunity to develop finer auditory discrimination and recognition skills. Subsequently, the students can experiment with the appropriate microphone distance for optimal sound quality.

With or without amplification, speakers and listeners can disrupt or distort the message. Students need practice to identify the behaviors characteristic of good speakers, such as facing the listener, speaking clearly, and indicating a beginning and an ending of a remark. Good listeners face the speaker, acknowledge the speaker's comments, and wait until the speaker is finished talking. Students also need to observe the characteristics of poor speakers and listeners. By using role playing in which the speaker's voice is amplified through the sound-field FM system, the whole class can participate in the evaluation of what behaviors the speaker and listener carried out appropriately or poorly. Of course, favorite parts chosen by students are always the poor speakers or listeners.

By modelling "unclear" messages we may teach the children the different factors that interfere with communication. Using the sound-field FM system as a simulation of communication on the telephone, the transmitter may be passed back and forth between two students. The teacher gives one student secret instructions about how to make his or her message more difficult to understand and the rest of the students have to guess how the student is distorting the message.

The sound-field FM system may be used in any number of comprehension activities to enhance the speaker's voice. Use of the FM transmitter during storytelling, children's theater creations, and singing provides better amplification of the children's voices to each other, resulting in more optimal communication.

## DEVELOPMENT OF LISTENING
## SKILLS IN THE CLASSROOM

Sound is an invisible characteristic of classroom conditions. Activities that encourage teachers and students to respond to, organize, interpret, and evaluate sound increase the visibility of sound. Sustained visibility is the most important factor. Rather than complete many of the suggested activities during the first few weeks after the sound-field FM system has been introduced, it is better to introduce one activity per week, and then extend the focus on listening strategies throughout the year. For example, during the first week, the teacher may work on microphone technique with the students. In the second week, the students may be asked to measure specific noises that occur in the classroom. The following week, the teacher may ask the students to bring in some socks or tennis balls for the chair legs. Later, the students may do a science experiment measuring various noises outside the school. The continued focus on listening activities and sound is a constant reminder of the importance of hearing, and specifically of the importance of high quality speech input.

**Follow-up is an essential part of any activity in order for the skill to become part of classroom routines and the children's listening practices**. Have the children create posters to be placed in the room to remind them of various lessons, such as the characteristics of good speakers and listeners. For the younger children, puppets can remind them of what good speakers and listeners should do. When someone in the class is not listening well, the teacher can ask the other students to suggest ways in which the poor listener might improve. "Noise monitors" build in awareness of noise levels in the class on an ongoing basis. Encouraging students to acknowledge and search for possible reasons for listening difficulties during daily activities focuses attention on ways to change the listening environment.


## MAXIMIZING THE USE OF THE SOUND-FIELD SYSTEM IN THE CLASSROOM

Encourage the students to use the sound-field FM system during any sharing time, announcements, or presentations to the class. This is generally not difficult since children of any age like to talk into a microphone. Teachers may amplify any of the audio or audiovisual sound sources such as record players, tape recorders, film projectors, and videocassette recorders, either by placing the microphone beside the sound source, or by patching directly into the transmitter, which is possible on some systems. All other teachers working with the class such as the French, music, or art teachers should receive a complete inservice training so that they, too, can use the sound-field FM system effectively.

**Table 11–2.** Key points in chapter.

| Key Points |
| --- |
| • Children are bombarded with noise during daily activities in the classroom. |
| • Only when noise is removed do the children become aware of its presence. Activities that contrast presence and absence of noise are necessary to highlight children's awareness of noise and its potential interference with communication. |
| • Once children become aware of various noise sources within and outside the classroom, listening activities can introduce ways to experiment with reduction of noise levels. Changing the ownership for noise control in the classroom from the teacher to the children is beneficial for creating lasting changes in noise levels. |
| • The sound-field FM system is useful for demonstration of speaker-listener strategies, and for the development of communication skills among students. |
| • Follow-up is essential to ensure carry-over of listening skills into everyday activities. |

## SUMMARY

When sound-field FM technology is introduced simultaneously with a focus on the development of listening skills and strategies, acoustic enhancement of the signal will be optimized. **A "one shot" approach to listening programming will *not* make a difference. Only the systematic inclusion of listening into everyday activities and ongoing explorations into various aspects of listening will enhance the auditory development of children in the classroom**. Key points of this chapter are presented in Table 11–2.

## RECOMMENDED RESOURCES FOR THE CLASSROOM TEACHER

Graser, N.S. (1992). *125 Ways to be a better listener.* East Moline, IL: Linguisystems Inc.

Kaner, E. (1991). *Sound science.* Toronto, Ontario: Kids Can Press Ltd.

Micallef, M. (1984). *Listening: A basic connection.* Carthage, IL: Good Apple Inc.

## REFERENCES

Boothroyd, A. (1994). *Hearing aids.* Paper presented at Technology for Communication and Education Symposium, Rochester, New York

Early Childhood Curriculum Committee. (1978). *Listening and speaking.* Adelaide, Australia: Publications Branch: Education Department of South Australia.

Edwards, C. (1991). Assessment and management of listening skills in school aged children. In C. Flexer (Ed.), *Seminars in hearing: Current audiologic issues in the educational management of children with hearing loss, 12*(4), 305–316. New York: Thieme Medical Publishers.

Erber, N. (1977). Evaluating speech perception ability in hearing impaired children. In F. Bess (Ed.), *Childhood deafness: Causation, assessment, and management.* New York: Grune & Stratton.

Finitzo-Hieber, T., & Tillman, T. (1978). Room acoustics effects on monosyllabic word discrimination ability of normal and hearing impaired children. *Journal of Speech and Hearing Research, 21,* 440–458.

Hetu, R., Truchon-Gagnon, C., & Bilodeau, S. (1990). Problems of noise in school settings: A review of the literature and the results of an exploratory study. *Journal of Speech Language Pathology and Audiology, 14*(3), 31–39.

Melancon, L., Truchon-Gagnon, C., & Hodgson, M. (1990). *Architectural strategies to avoid noise problems in child care centres.* Montreal, Canada: Groupe d'acoustique de l'Universite de Montreal.

Neuman, A., & Hochberg, I. (1983). Children's perception of speech in reverberation. *Journal of the Acoustical Society of America. 73,* 2145–2149.

Yacullo, W., & Hawkins, D. (1987). Speech recognition in noise and reverberation by school-age children. *Audiology, 26,* 235–246.

# CHAPTER
# 12

# MARKETING SOUND-FIELD AMPLIFICATION SYSTEMS

*Laurie A. Allen*
*Karen L. Anderson*

The concept of sound-field amplification has been promoted since the MARRS Study first became a U.S. Department of Education National Diffusion Network (NDN) project in 1979. The efforts of this NDN project have been primarily aimed at educators as a way to improve educational effectiveness for students with undetected hearing impairment and/or other children with attentional or auditory learning problems. Only in approximately the last five years have audiologists become actively interested in sound-field amplification use for students with recurrent otitis media as well as for students with central auditory processing disorders, or for those with unilateral, high-frequency, or mild sensorineural hearing loss (see Chapter 4). This increased interest has precipitated the creation of several brands of classroom sound-field amplification systems that are now available from different manufacturers in a wide price range.

As has been documented in previous chapters, the use of sound-field classroom amplification is consistently beneficial to students and teachers. Despite this finding, not every principal or superintendent welcomes a piece of equipment that teachers will fight to keep once they try it in their classrooms. In other words, the very success and proliferation of sound-field amplification use in the classroom is often perceived as a threat to budget-conscious school administrators. This chapter will thus examine how sound-field amplification can be promoted effectively in the schools.

## PREPARATION

Preparation represents the key to the successful marketing of sound-field classroom amplification equipment. It is much like the symphony musician who practices the music and attends rehearsals in preparation for the big concert. All of his or her efforts will be in vain if he or she plays too soon and destroys the effect of the overture. It is important to toot your own horn, but don't come in before the down beat!

The leg work must be done first in order to achieve success with the use of classroom amplification equipment. The following steps must be followed:

- A receptive school must be found with an administrator who is excited about trying out the equipment.
- A teacher will need to be selected who wants to try the equipment and who is comfortable using electrical equipment.
- The right equipment must be selected to meet the physical needs of the classroom.
- The teacher must receive thorough inservice training so that she or he feels comfortable using the equipment. Then, and only then, can you start marketing the equipment. If a breakdown occurs in any of the above areas, it will slow down your progress.

## GETTING STARTED—THE PILOT PROJECT

In every school district, there are usually a few school principals who thrive on trying innovative programs or teaching methods in their building. In addition, there is often active parent participation and community support visible in these schools and they are the schools that are usually highlighted in the local newspapers. This newspaper coverage is an indication that there is a school principal that likes to be on top of things and who wants the public to know of all of the happenings in their building. This type of administrator represents a good candidate to approach regarding the use of sound-field classroom amplification. It is important to sit down with this person and describe the use and benefits of the amplification equipment. It is helpful to provide a written summary of the sound-field amplification concept and benefits (Appendix A). If possible, provide the principal with the name and telephone number of another administrator who has had experience with the equipment. Would the principal be willing to try this type of equipment in one of the classrooms? Would it be feasible to study single or group student performance/achievement in several rooms to compare the affects in classrooms both with and without the amplification (parent permission requirements, test timelines, etc.)? Brainstorm how the principal could see trying the sound-field amplification concept in his/her school.

## TEACHER SELECTION

Once you have found a receptive school administrator, it is important to find a teacher who is willing to try the equipment. Since classroom size, shape, and acoustic treatment can vary from room to room it is always beneficial to have several candidates to help facilitate the matching of the right equipment with the teachers and the needs of their classrooms.

Ask the principal to arrange an inservice meeting with the teachers, so that you can give them a brief overview of the equipment and why you would like to try it in their building. This will provide an excellent opportunity to assess which teachers really seem interested in trying out the equipment; then ask for volunteers. With a list of volunteers in hand, take a quick tour of the building to visit the classrooms and form a better idea about which rooms would benefit the most from the use of the equipment. An exceptionally large room will be more needy than a small one. This first-hand information will ease the selection process.

Points to consider when selecting a teacher include:

- Select a teacher who wants to try out the equipment. Avoid forcing it on anyone. If the teacher tells you that she or he has a really strong voice that carries well, then back off and try another teacher. If the principal suggests Teacher A and you find the latter is not receptive to the idea go back to the principal and diplomatically suggest other possible candidates. Perhaps the room with high noise levels (e.g., buzzing lights) or Teacher B with the really soft voice would work better. This way the principal will learn that the equipment can be used in a variety of classrooms and will still have a say in the teacher selection process.

- Look for a teacher who is articulate and respected by his/ her peers. The sound-field amplification equipment, if installed and maintained properly, will sell the benefits of improved classroom listening by itself. A first-class teacher can do much to speed the promotion process along at both the school district and community levels.
- Select a teacher who welcomes visitors to the classroom. Once you have the amplification system installed you want others, such as the school superintendent, PTA president, newspaper reporter, and so on, to come and experience the difference that the amplification can make in the classroom setting.
- Select a teacher who is willing to provide you with feedback. The teacher needs to know that you will be asking that feedback questionnaires such as the Screening Instrument For Targeting Educational Risk (SIFTER) (Anderson, 1989) need to be completed and returned to you as part of the trial period (see Appendix B).

## EQUIPMENT SELECTION

Several companies manufacture sound-field classroom amplification systems. These systems vary in features, price, and user friendliness. The easiest way to obtain equipment for short-term use is to contact the manufacturers directly. Often they will allow demonstration units to be placed in a school for a 30-day trial period. Perhaps such trial periods can be lined up back-to-back so that a different brand of equipment can be tried by the same teacher in order to allow comparison of the different brands. Be aware that each individual system will require some additional teacher inservice training. For more manufacturer information refer to Chapters 1 and 9.

If time is short and it is not possible to carry out several trials of different equipment, then ask the manufacturers for names and telephone numbers of satisfied customers. Ask specifically for names of people who do more that just sell the equipment to schools. Someone like an educational audiologist or a teacher who has had the experience of installing and maintaining the equipment on a regular basis will provide valuable information.

## ROOM SELECTION AND INSTALLATION

Once you have equipment to use and a list of volunteer teachers, the room selection becomes much easier. With knowledge about the differences among brands of equipment, it should only take a few minutes to determine if amplification equipment can be successfully installed in any given room. Some questions to consider during the selection process include:

- Can all of the system's speakers be used in this room without having speaker wire draped across the travel paths in the room? Four speakers usually offer better coverage and less feedback problems than just one or two speakers.
- How many electrical outlets are available for use? Some systems require more than one outlet for the receiver/amplifier. Other brands require that each speaker be plugged into an outlet (then no speaker wires are used). Be sure to consult with the teacher to find out which outlets in the room are available for use.
- Are there places where speakers can be set or mounted so that they provide appropriate sound coverage throughout the room (e.g., one per wall rather than along one or two walls)? Counterspace is at a premium in most classrooms.
- Is there a location where the receiver or receiver/amplifier can be placed so that it is away from excessive heat, cold, sinks, aquariums, and computers?
- Can the equipment be installed by one person or will school district personnel (e.g., custodian, buildings and grounds electrician, etc.) need to be available to help?

Choosing a classroom where the amplification equipment will have the greatest impact is also an important consideration. The amplification will be in use only for large group instruction; thus avoid classrooms that have a small number of students who receive mostly one-to-one or small group instruction. Try to choose a room that has several factors that undermine the listening environment, such as the teacher with the exceptionally soft voice, the room with many students, a large number of English as a Second Language (ESL) students or a physically large or noisy classroom. Be sure to demonstrate to the teachers and any visitors the difference in the listening environment when the system is on and off.

## DATA COLLECTION

Once the parameters and personnel of the pilot project have been determined and any necessary parent permissions have been obtained, equipment installation and training can take place and data collection can begin. Be sure to allow ample time to train the teacher thoroughly on use of the equipment and stop by the classroom on a regular basis to troubleshoot any problems. At the end of the project it is important to have the involved principals and teachers complete an evaluation form (see Appendixes C and D). It is also important to gather feedback from several, if not all, of the students by way of an evaluation form (Appendix E). Student reading levels must be considered when the feedback form is designed. This subjective feedback information from your local school district teachers and student population is very meaningful to your school superintendent, PTA members, and local service organizations who may very well be your future sources of funding for equipment. This local information along with the current research literature makes for a very convincing case when you are presenting to others in your community.

Once data is collected and analyzed information may be disseminated. First and foremost, share the information in a brief way with the local school board members and Director/Coordinators of Special Education. Share your results with your State Superintendent of Public Instruction. Send a copy to your local principals.

Volunteer to talk at local service organizations and then ask if they want to adopt a school or a classroom and provide the amplification equipment needed. Volunteer to write up an article for the school newsletter or write up a separate handout the children may take home to their parents. Have stickers printed with "Amplification helps me hear better in school!" and pass them out to all of the students in the amplified rooms. Sport a large button on your lapel that says "Classroom amplification makes a difference!" This will generate interest and questions. Push yourself to spread the word because your efforts to promote the sound-field amplification concept will result in positive gains for many children.

## COST EFFECTIVENESS OF SOUND-FIELD CLASSROOM AMPLIFICATION

Anything that helps children learn in a cost-effective and positive way should be of interest to all educators and parents. However, the bottom line will always be: where will the school obtain the money to pay for sound-field amplification equipment? School administrators have difficulty when it is obvious that students and teachers are benefiting from the equipment and it is now wanted by all of the teachers but the funds to purchase the equipment are not available. The best way to address this problem is to show the administrators that over time they would more than likely save money for the district as a result of the use of the sound-field amplification equipment. Create a sample plan which includes a scenario for minimal and modest expansion. For example, if two classes per grade K-2 in a typical elementary school were amplified it would cost approximately "X" amount of money as opposed to amplifying all classes grade K-2 which would cost "Y." Compare this expenditure to some other recognized cost, such as carpeting or substitute pay when a teacher takes days off because of vocal abuse related maladies.

Use the growing body of available sound-field amplification research (see Chapter 8) to develop cost savings projections through decrease in special education needs (Educational Audiology Association, 1991). One excellent illustration of the impact of classroom amplification on special education comes from the Putnam County School District in Ohio (Phonic Ear, 1994). From 1985 to 1990, the district phased in 60 sound-field amplification systems to help identified students with learning disabilities hear more easily in the mainstreamed classroom setting. The cost of the amplification equipment at that time was approximately $1,500 per unit; however, current brands are available at a significantly lower cost. The number of students placed in learning disabilities (LD) programs has since declined nearly 40 percent (26 students) during those five years. During 1990, the cost per year of placing a child in special education in Ohio was $2,600 (Berg, 1993). To continue the scenario into future years, we can predict that if Putnam County experiences another decline of 26 students placed in LD programs over another 5-year period, the minimum savings at the end of the period would be:

| | |
|---|---|
| Cost of savings per year per potential LD student | $2,600.00 |
| Number of school years not requiring special programming | ×5 |
| Total savings per student (keeping special education costs constant) | $13,000.00 |
| Number of students not needing special education assistance because of amplification intervention (40% decline) | _____ ×26 |
| Minimum savings over five-year time period = | $338,000.00 |

## FUNDING SOUND-FIELD AMPLIFICATION EQUIPMENT

Classroom amplification equipment is relatively inexpensive ($700–$1,800) and benefits both the students and the teachers. With classroom amplification providing such a broad educational and personal impact numerous sources may be approached for financial assistance. If several amplification systems are desired it would probably be best to write a grant to fund the equipment. As part of the National Diffusion Network, the MARRS project is an excellent source for obtaining the first few amplification systems in your district. It is uncertain how long the MARRS project will remain an NDN project (it is now in its third 5-year funding cycle) but funds are available as long as it remains in the pool of NDN projects. Typically, NDN projects are funded to a maximum of $5,000 (amount may vary) and must be used for a specified project such as handicapped preschoolers, ESL students, and so on. Your school district should be able to tell you how to locate the State NDN Project Facilitator in your area or you may contact the National Diffusion Network Division, Department of Education, 1200 19th St. N.W., Room 714F, Washington, DC, 20208 Telephone (202) 653-7003. The NDN grant application form is brief and relatively easy to complete. Applications are taken only in the spring with funds being available for equipment purchase in the fall. If manufacturers are aware that you have been chosen to receive grant monies but, for example, will not actually receive the dollars until November, they will often allow you to obtain the equipment when school starts and arrange for you to pay when the dollars are released to the district.

Many other grant sources exist but the competition between those applying for projects is often fierce and funding for the purchase of equipment is limited. In general, it is easier to find funding for an amount to support a small pilot project (i.e., under $10,000) than for a large, district-wide project. Even $5,000 or $10,000 is enough to fund a well organized pilot project adequately. If there is a person in charge of grantwriting for your school district contact them for advice. Grant sources can be found in reference books at your public and college libraries. It may be worthwhile to write up a letter that contains 1–2 paragraphs describing your intended project (including time line, district involvement, predicted outcome, and cost) and send the letter to many potential grant sources with a request for an application form if the project appears to be within their funding parameters. It may take a year to obtain funding but if you inquire to a large number of potential sources it is likely that several will be interested in your project and that funding will be received from at least one.

Another funding source can be the schools' PTA. Equipment that has a direct positive impact on their children and their teachers has a very strong appeal to parent groups. Schools typically have fund-raising activities throughout the year. Talking with the parent-teacher group, formalizing a plan of action, and advertising that some of the money will go to the purchase of amplification equipment provides a real incentive for parents to contribute. If you consider that the average cost of an amplification system is about $1,000 ($550–$1,800), then that can equate to profits from two book fairs or four school bake sales for a school of 400–500 students. Perhaps a percentage of the profits from the lounge concession machines or from the school supply store could be earmarked for buying the amplification equipment. Finding a funding source is an opportunity for creativity!

If the school you wish to amplify does not have an active PTA consider speaking to local service organizations (e.g., SerToMa, Lions Club, Rotary Club, etc.) or determine if the school has a business or corporate sponsor which would like to "adopt" a classroom or grade. Corporate sponsors may be willing to support a relatively low-cost, short-term, highly visible project. Even if only one system was purchased each year there would be an amplified classroom at each grade level in seven years at a typical elementary school. Any assistance should be reported to the local newspaper and perhaps a small plaque could be mounted on the equipment or outside the classroom stating the name of the organization or sponsor which provided the amplification equipment for the classroom. Contact people for the service organizations may be identified by calling your local Chamber of Commerce or Information and Referral office.

## THE SINGLE STUDENT APPROACH

As an alternative to the pilot project approach to promotion it is sometimes feasible to introduce sound-field amplification into a school building by obtaining it for a student with recognized hearing needs. It is possible to approach the use of the equipment for one student as an opportunity for the whole school staff to learn more about hearing loss and how sound-field amplification can benefit students and teachers. It must be acknowledged, however, that if the equipment does not markedly benefit the student it will be removed from the classroom even it the teacher loves it and its benefit are evident to the rest of the students. Inviting representatives from the parent-teacher organization to observe in the classroom and for them then to consider funding single units for use in the classrooms may be fruitful.

For the single student approach, a likely candidate must be chosen. Choose your student candidates carefully—too many recommendations to any one principal or administration may cause your opinion to no longer be considered credible. Even in well funded school districts there are only so many dollars budgeted for anticipated equipment needs for special education students. With careful introduction to the use and benefits of this equipment you can develop of a vocal group of amplification supporters (teachers, principals, parents) and the budget for amplification equipment may increase slowly from year to year.

Our purpose in providing classroom amplification is to improve the students' listening environment so that they can learn as "normally" as possible. If the student is already perceived as a normal learner it is difficult to be successful in promoting the amplification equipment to help him or her hear, and subsequently, learn better. Functional abilities such as student attention and class participation should be discussed as areas of need as well as potential improvement in academic performance. To illustrate the level of difficulty a child is experiencing, obtain information from the classroom teacher about the child's performance prior to promoting classroom amplification. This information can be obtained by asking the teacher to complete a checklist like the Screening Instrument For Targeting Educational Risk (SIFTER, see Appendix B).

By working with the teacher to discuss the SIFTER results and describing sound-field amplification equipment, it is possible to discover if the teacher would be willing, or preferably excited, to try sound-field amplification for the benefit of the student and the class. If the teacher is not a willing participant during the trial period the chances of the trial being successful and inciting excitement about the sound amplification concept throughout the school is doubtful at best. Also, if the principal is not a strong supporter of the trial period and is not interested in the results of sound-field amplification for all students, your efforts probably will not be successful. As with the pilot project approach, preparation is the key to success!

# REFERENCES

Anderson, K. (1989). *Screening instrument for targeting educational risk (SIFTER).* Seattle, WA: Educational Audiology Association.

Berg, F. (1993). *Acoustics and sound systems in schools.* San Diego: Singular Publishing Group Inc.

Educational Audiology Association. (1991). *Sound field classroom amplification: A collection of writings by members of the Educational Audiology Association.* Seattle, WA: Educational Audiology Society.

Phonic Ear. (1994). *Facts, figures & FM.* Petaluma CA.

# APPENDIX A
# THE USE OF SOUND FIELD AMPLIFICATION OF THE TEACHER'S VOICE IN THE REGULAR EDUCATION CLASSROOM—A SUMMARY OF STUDIES

## The MARRS Project: Mainstream Amplification Resource Room Study

**THE CONCEPT:** MARRS is an National Diffusion Network (NDN) project that uses a wireless FM microphone system for soundfield amplification of the classroom teacher's voice in order to enhance oral instruction, lessen teacher voice fatigue, and improve student academic achievement. Amplification of the teacher's voice above background noise is provided to all students in the classroom so that those in the back row can hear as clearly as those in the front of the class.

**THE INTENDED POPULATION:** The MARRS Project was primarily intended as a means of helping students with mild or minimal fluctuating hearing losses compensate for poor classroom acoustics enabling them to remain in the mainstream without expensive referral and identification procedures. Data obtained by MARRS Project staff has revealed that 20–25% or more of the current school population have academic difficulties co-existing with minimal hearing loss (defined as 15–40 dB). If episodes of this degree of hearing loss are frequent, children can miss significant language experience and academic instruction which can cause them to develop learning difficulties that may subsequently require special education services. Education benefits of improved classroom listening via amplification of the teacher's voice have been repeatedly illustrated for thousands of students with normal hearing as well as those students with hearing loss.

**SUMMARY OF BENEFITS:** The quality of oral instruction is enhanced with amplification since all children receive a clear audible instructional signal throughout the classroom, regardless of interfering noise and where they are seated. Teachers using amplification report improved student attention, fewer distractions, and less need to repeat instructions. Classroom management is enhanced and discipline problems are diminished because the teacher has better voice-control of every student in the classroom. Almost all students comment the amplified voice helps them pay attention, better understand oral directions, shut out distracting noises and hear the teacher without straining. The evidence for improved teaching and quality of instruction is reflected in the statistically significant gains in reading and language achievement test scores for K-6 students included in classrooms using amplification (students with and without hearing loss). These improvements were evident after only one year of use and the improved academic scores have been maintained for as much as 3 years. The amplification was found to be more cost effective than supplementary resource room instruction in: 1) staff utilization (requiring fewer personnel to achieve the same or superior academic growth), 2) lower initial and continuing educational costs and 3) personal costs to students who avoided the stigma, segregation and restrictions of special placement.

## RESEARCH RESULTS—in brief:

- In 1981, at the end of the original 3 year MARRS study, data was analyzed for three treatments of identified target students. The MARRS study found that approximately 30% of children in grades 3–6 failed a 15 dB hearing screen. These target students were divided into three groups: 1) typical classroom settings, 2) regular classroom instruction with supplemental resource room instruction, and, 3) regular classroom instruction with sound field amplification of the teacher's voice. Amplification of the teacher's voice resulted in significant improvement (>.05 level) in academic achievement test scores of the minimal hearing loss students. These gains were achieved at a faster rate, to a higher level at a lesser cost than gains achieved by students in the more traditional resource room model typically utilized for students requiring special help. Increases in test scores ranged from 1/3 or over 1 standard deviation. The significant increases were not observed in scores of students in the resource room in the same time interval. The significance of the findings in the MARRS study is that for some students with minimal hearing losses, significant educational instruction effects can be achieved by sound field amplification. Furthermore, these gains can be cost effectively realized within the regular classroom without the need for stigmatizing labeling and segregation as well as expense and scheduling complications of special class placement. National Diffusion Network recognition of the MARRS Project as an exemplary educational program was granted in 1981.

- In 1982, a study investigated the effects of sound field amplification on the test taking performance of 131 second and third grade children having either a minimal hearing loss or normal hearing. Also, a behavior rating scale completed by the classroom teacher was correlated with the two groups. The results indicated improved performance on a dictated spelling test for students having minimal hearing loss. The behavior scale had a negative correlation, indicating students with a minimal hearing loss were viewed by the classroom teacher as impulsive, overactive and having a weak attention span.

• In 1983, the listening ability of kindergarten students under close, distant and sound field amplification was explored. Students with minimal hearing loss (15-40 dB) were identified. A high fidelity tape recorder was placed on the teacher's desk and presented words that students were to identify from 25 multiple choice items by placing an X on the appropriate picture. When the teacher's desk was at the front of the room (average of 12 feet from students), the normal hearing children achieved an average listening accuracy of 91% and the children with hearing loss had an average score of 81%. With the teacher's desk in the center of the classroom (average of 6 feet from the students), scores improved to 98% and 96% for the normal and hearing loss groups respectively. When words were presented via sound field amplification, scores for both groups averaged 98%. Reported results indicated: 1) students with minimal hearing loss do not listen as effectively as normal hearing students from a distance in a kindergarten classroom with "good" acoustical conditions, and 2) listening problems may be alleviated either by a teacher moving up close to students or using amplification equipment.

• In 1986, data were gathered from 40 public school classrooms in grades K-6 from sites in Illinois, Kentucky, Minnesota and Missouri. Students with minimal hearing loss were identified at each site and placed at random in amplified classrooms or control (non-amplified) classrooms. Data from each site contained the pre-test and post-test for each target student from the amplified and control groups. It was found that students with minimal hearing loss who learn in an amplified environment perform significantly better on standardized achievement tests of reading and language (P<.05). This was true across the site, grade, and measurement scale used. Subjective results revealed 85–90% of teachers found the system beneficial to themselves and to their students for the improvement of teaching and quality of instruction. Eighty-five to ninety percent of students judged the system to be beneficial to them in improving their ability to hear and to help in their schoolwork and 90% of administrators surveyed had positive responses to questions assessing the effectiveness of amplification use.

- In 1990, after 60 sound field amplification units had been phased in over a 5 year period it was found that the number of students placed in LD programs had declined nearly 40%. Also, findings indicated that 43% of students had minimal hearing loss on any given day and approximately 75% of primary-level children attending LD classes also did not have normal hearing. Using the Iowa TBS to evaluate achievement the following was noted: the amplified Kindergarten showed the most dramatic results with significantly higher scores on listening, language, and word analysis. The amplified Grade 1 classes showed superior performance on word analysis and vocabulary. The amplified Grade 2 classrooms showed better scores on math concepts and computation, and the amplified Grade 3 classrooms showed superiority on math computation concepts and reading. Formal classroom observations indicated that students in amplified classrooms had better student production and on-task behaviors. Principals also noticed fewer teacher absences due to fatigue and laryngitis.
- In 1990, children with developmental disabilities in a primary-level class utilizing sound field amplification made significantly fewer errors on a word identification task than they made without amplification. Children were observed to be more relaxed and responded more quickly in the amplified condition.
- In 1993, children with ongoing hearing loss and histories of chronic ear problems were identified in 12 classrooms. The teachers, who were unaware of the target students, completed S.I.F.T.E.R. educational screening forms for all students in 6 amplified and 6 unamplified classrooms. Results showed that about one-third of the children have early and continuing hearing problems. Most important, children benefit from classroom amplification, whether or not they have hearing problems.
- In 1994, the listening abilities of children who have learned English as a second language were studied under amplified and non-amplified conditions. Results indicated that ESL students experienced significant difficulty understanding spoken English in a typically noisy classroom environment. Significant improvement in understanding ability of the ESL students was revealed under amplified classroom conditions.

# REFERENCES

1) Sarff, L., Ray, H., & Bagwell, C. (1981). Why not Amplification in every class-room? *Hearing Aid Journal*, October.

2) Burgener, G. & Deichmann, J. (1982). Voice amplification and its effects on test taking performance. *Hearing Instruments*, 33:11.

3) Jones, J. & Berg, F. (1983). Listening of kindergarten students under close, dis-tant, and sound field FM amplification conditions. Unpublished Master's the-sis. Logan, UT: Utah State University.

4) Ray, H. (1992). Revalidation Submission: Project MARRS. WOVSED, P.O. Box E, Norris City, IL 62869. Resubmitted to National Diffusion Network.

5) Flexer, C. (1989). Turn on sound: an odyssey of sound field amplification. *Edu-cational Audiology Association Newsletter*, 5:5.

6) Flexer, C., Millin, J. & Brown, L. (1990). Children with developmental disabili-ties: effect of sound field amplification on word identification. *Language Speech & Hearing Services in Schools, 21.*

7) Flexer, C., Richards, C. & Buie, C. (1993). Sound field amplification for regular K and 1st grade classrooms. Poster Session at the annual meeting of the American Academy of Audiology in Phoenix AZ.

8) Crandell, C. (1994). Use of sound field amplification with ESL students. Pre-sented at the American Academy of Audiology annual meeting, Richmond VA.

# APPENDIX B
# S.I.F.T.E.R.
# SCREENING INSTRUMENT FOR TARGETING
# EDUCATIONAL RISK

### By Karen L. Anderson, Ed.S., CCC-A

STUDENT _____     TEACHER _____     GRADE _____

DATE COMPLETED _____     SCHOOL _____     DISTRICT _____

The above child is suspect for hearing problems which may or may not be affecting his/her school performance. This rating scale has been designed to sift out students who are educationally at risk possibly as a result of hearing problems.

Based on your knowledge from observations of this student, circle the number best representing his/her behavior. After answering the questions, please record any comments about the student in the space provided on the reverse side.

| | | | |
|---|---|---|---|
| 1. What is your estimate of the student's class standing in comparison of that of his/her classmates? | UPPER<br>5   4 | MIDDLE<br>3   2 | LOWER<br>1 |
| 2. How does the student's achievement compare to your estimation of her/his potential? | EQUAL<br>5   4 | LOWER<br>3   2 | MUCH LOWER<br>1 |
| 3. What is the student's reading level, reading ability group or reading readiness group in the classroom (e.g., a student with average reading ability performs in the middle group)? | UPPER<br>5   4 | MIDDLE<br>3   2 | LOWER<br>1 |

ACADEMICS

| | | | |
|---|---|---|---|
| 4. How distractible is the student in comparison to his/her classmates? | NOT VERY<br>5   4 | AVERAGE<br>3   2 | VERY<br>1 |
| 5. What is the student's attention span in comparison to that of his/her classmates? | LONGER<br>5   4 | AVERAGE<br>3   2 | SHORTER<br>1 |
| 6. How often does the student hesitate or become confused when responding to oral directions (e.g., "Turn to page . . .")? | NEVER<br>5   4 | OCCASIONALLY<br>3   2 | FREQUENTLY<br>1 |

ATTENTION

| | | | |
|---|---|---|---|
| 7. How does the student's comprehension compare to the average understanding ability of her/his classmates? | ABOVE<br>5   4 | AVERAGE<br>3   2 | BELOW<br>1 |
| 8. How does the student's vocabulary and word usage skills compare with those of other students in his/her age group? | ABOVE<br>5   4 | AVERAGE<br>3   2 | BELOW<br>1 |
| 9. How proficient is the student at telling a story or relating happenings from home when compared to classmates? | ABOVE<br>5   4 | AVERAGE<br>3   2 | BELOW<br>1 |

COMMUNICATION

| | | | |
|---|---|---|---|
| 10. How often does the student volunteer information to class discussions or in answer to teacher questions? | FREQUENTLY<br>5   4 | OCCASIONALLY<br>3   2 | NEVER<br>1 |
| 11. With what frequency does the student complete his/her class and homework assignments within the time allocated? | ALWAYS<br>5   4 | USUALLY<br>3   2 | SELDOM<br>1 |
| 12. After instruction, does the student have difficulty starting to work (looks at other students working or asks for help)? | NEVER<br>5   4 | OCCASIONALLY<br>3   2 | FREQUENTLY<br>1 |

CLASS PARTICIPATION

| | | | |
|---|---|---|---|
| 13. Does the student demonstrate any behaviors that seem unusual or inappropriate when compared to other students? | NEVER<br>5   4 | OCCASIONALLY<br>3   2 | FREQUENTLY<br>1 |
| 14. Does the student become frustrated easily, sometimes to the point of losing emotional control? | NEVER<br>5   4 | OCCASIONALLY<br>3   2 | FREQUENTLY<br>1 |
| 15. In general, how would you rank the student's relationship with peers (ability to get along with others)? | GOOD<br>5   4 | AVERAGE<br>3   2 | POOR<br>1 |

SCHOOL BEHAVIOR

Copyright ©1989 by Karen Anderson

**Reproduction of this form, in whole or in part, is strictly prohibited.**

Additional copies of this form are available in pads of 100 each from
The Educational Audiology Association Products Manager
Donna Fisher, P.O. Box 3811, Little Rock, AR 72203
ISBN 0-8134-2845-9

## TEACHER COMMENTS

Has this child repeated a grade, had frequent absences or experienced health problems (including ear infections and colds)? Has the student received, or is he/she now receiving, special support services? Does the child have any other health problems that may be pertinent to his/her educational functioning?

## THE S.I.F.T.E.R. IS A SCREENING TOOL ONLY

Any student failing this screening in a content area as determined on the scoring grid below should be considered for further assessment, depending on his/her individual needs as per school district criteria. For example, failing in the Academics area suggests an educational assessment, in the Communication area a speech-language assessment, and in the School Behavior area an assessment by a psychologist or a social worker. Failing in the Attention and/or Class Participation area in combination with other areas may suggest an evaluatin by an educational audiologist. Children placed in the marginal area are at risk for failing and should be monitored or considered for assessment depending upon additional information.

## SCORING

Sum the responses to the three questions in each content area and record in teh appropriate box on the reverse side and under Total Score below. Place an X on the number that corresponds most closely with the content area score (e.g., if a teacher circled 3, 4 and 2 for the questions in the Academics area, an X would be placed on the number 9 across from the Academics content area). Connect the X's to make a profile.

| CONTENT AREA | TOTAL SCORE | PASS | MARGINAL | FAIL |
|---|---|---|---|---|
| ACADEMICS | | 15 14 13 12 11 10 | 9  8 | 7 6 5 4 3 |
| ATTENTION | | 15 14 13 12 11 10 9 | 8  7 | 6 5 4 3 |
| COMMUNICATION | | 15  14  13  12  11 | 10 9 8 | 7 6 5 4 3 |
| CLASS PARTICIPATION | | 15 14 13 12 11 10 9 | 8  7 | 6 5 4 3 |
| SOCIAL BEHAVIOR | | 15 14 13 12 11 10 | 9 8 | 7 6 5 4 3 |

# APPENDIX C
# SOUND FIELD CLASSROOM AMPLIFICATION
# TEACHER QUESTIONNAIRE

School District ____EXAMPLE____ School_____

Grade Taught _2_ Sex _F_ Years of Teaching Experience __17__

Date Installed _____ Date Questionnaire Completed _____

1. During what subjects do you use classroom amplification? Reading, math, writing, spelling, language, science, social studies

2. Approximately how many hours do you use amplification per day? 3 hours

3. Number of students in amplified classroom? 24

4. How were you selected for participation? Personal interest _____
   Volunteer __X__ Administrative Assignment _____ Other _____

5. Describe you natural unamplified voice:
   Pitch = Low ___ Medium _X_ High ___
   Loudness = Soft __X_ Average _X_ Loud ___

6. Do you feel the classroom amplification system has helped students? Why or why not? Yes, it was amazing the difference in student attention with amplification. They were *all* paying attention and participating rather than the front few students.

7. Do you feel that classroom amplification has been particularly helpful for any specific subject(s) or classroom assignments? Describe: I like it for a wide variety of subjects but especially discussion-oriented lessons (social studies, science, reading). I tried to repeat the students responses so everyone was aware of the responses. Also, for students reading their stories and writing assignments. Prior to the amplification students paid very little attention to classmate responses and stories.

8. Has the amplification been helpful to you personally in classroom instruction? In what ways? Yes, I found I could speak "softly" and get more attention and be more dramatic and expressive - without speaking everything at the same loudness. Also, I could talk while I wrote on the board which was a terrific time saver!

9. How do you feel about using amplification? Has your attitude changed since you first started using the system? Yes, at first I liked the novelty, but now I think it's more of a necessity. If I hadn't tried it I never would have realized what a difference it makes!

## APPENDIX D
## SOUND-FIELD CLASSROOM AMPLIFICATION
## QUESTIONNAIRE (ADMINISTRATOR)

School District _____EXAMPLE_____ Date_____

Building _____ Number of Amplified Classrooms __4__

Date Installed _____ Date Questionnaire Completed _____

1. Do you think the classroom amplification system has been helpful to students? Comments: YES. I feel the amplification system has provided a number of positive factors that have greatly influenced the learning rate or progress of students. It reduces repetition of instructions from the teacher as well as keeping the students more attentive toward instruction.

2. Do you think the amplification system has been helpful to teachers? Comments: Very much so! The results and positive attitudes of our staff have been readily observable. Voice fatigue is virtually eliminated and the convenience it provides to staff who can face the blackboard, etc. without worrying IF a student can hearing him/her is very evident.

3. How do you feel about the use of classroom amplification? Comments: Very Positive and sold on its value toward serving students NOT with just minimal hearing loss but all students within the classroom setting.

4. What observations have you had from parents or the community? Comments: We have had very positive feedback from parents, community leaders, students, other educational agencies and from staff members. Our local newspaper had a front page 4 column story on its utilization within our school.

5. What have been other staff members comments about the classroom amplification system(s)? They have been very excited about being able to audibly teach all students.

6. What has been you experience with regard to servicability, repair, maintenance of equipment? Comments: So far no notable problems obtaining service. Equipment has required minimal care or servicing.

7. What other comments would you care to make? I feel it has been of *great* value to our students and has provided a *great* service or teaching tool to each faculty member utilizing the amplification equipment.

# APPENDIX E
# SOUND-FIELD CLASSROOM
# AMPLIFICATION STUDENT QUESTIONNAIRE

School _____EXAMPLE_____    Teacher _____

Grade _4_ How long has your teacher used the microphone? _6 weeks_

1. Do you think amplification of the teacher's voice has helped you in your school work?

       YES _____        NO _____

2. Please tell how it has helped.

    I could hear better.

    It got my attention.

3. Would you like to see amplification used in more classrooms?

       YES __X__        NO _____

4. If you could change the amplified system in some way, how would you change it?

      NONE!    I like it!

# INDEX